MODERN PROSPECTING

HOW TO FIND, CLAIM AND SELL MINERAL DEPOSITS

By Roger McPherson

Gem Guides Book Co.
315 Cloverleaf Dr., Suite F
Baldwin Park, CA 91706

Library of Congress Control Number: 00-136492
ISBN 1-889786-16-0

Cover: Scott Roberts
Editor: Janet Francisco

NOTE:
Due to the possibility of personal error, typographical error, misinterpretation of information, and the many changes due to man or nature, *Modern Prospecting: How to Find, Claim and Sell Mineral Deposits*, its publisher, and all other persons directly or indirectly associated with this publication, assume no responsibility for accidents, injury, or any losses by individuals or groups using this publication. In rough terrain and hazardous areas, all persons are advised to be aware of possible changes due to man or nature that occur.

Every effort has been made to ensure accuracy of the information in this book. All contents are based on data available at the time of publication. However, time brings change and the publisher can accept no responsibility for any loss or inconvenience sustained by any visitor as a result of information contained in the book.

Readers are invited to write to the publisher with their comments, suggestions, and ideas.

TABLE OF CONTENTS

ACKNOWLEDGEMENTS

I would like to thank my prospecting teacher, James Madonna, for introducing me to this subject through his University of Alaska Mining Extension classes, laboratory work, and field trips. University of Alaska geology professor Rainer Newberry, geologists with Kinross Gold Company, the Geological Society of Nevada, State of Alaska geologists with the Division of Geological and Geophysical Surveys, and many other individuals have encouraged me and freely shared their expertise.

Sharon Kessey's careful editing has greatly improved the readability of these chapters. She also puts up with my annual pronouncement that this season I am certain to find my bonanza deposit.

Although I carefully researched this material, and others have critiqued some chapters, this is ultimately a subjective account of my perceptions as a prospector.

INTRODUCTION

Prospecting for gold and other minerals is a great way to be outdoors, keep in shape, explore the complexity of the geological record, and maybe even strike it rich. Many of the stories included here explain how I learned effective prospecting, often missing the "big one" in the process. Beyond gold fever lies a systematic approach to mineral exploration accessible to everyone.

With just a gold pan, a couple of screens, a small shovel, a rock hammer, a hand lens, and a few sample bags anyone can start prospecting. But, to have a better chance of finding a deposit, it's necessary to identify a favorable area to prospect, do some research on it, follow through with thorough reconnaissance and testing, and recognize the signposts. There are still bonanzas out there waiting to be discovered by people equipped with these basic skills.

In spite of taking good basic prospecting classes and living in a wide open mining district, I overlooked mineralized areas and staked faulty claims. I drilled when I should have walked away, and I spent more time and money on a property than it deserved. As driller Rocky MacDonald advised me, "There's an order to things. When you have a raw prospect, do all the steps...." From research to soil sampling to drilling, I've outlined the procedures to follow in order to reduce the expenses and the trial-and-error learning time.

Most of these techniques have been developed since the 1950s. Today we are able to look deeper into the earth, measure element amounts in rock and soil samples to parts per million or even parts per billion, and understand how regional tectonic trends relate to mineral deposits. Using these new tools, an individual prospector can compete with anyone in the field.

Early on I realized the importance of assays, so I visited assay laboratories and investigated their procedures to demystify this subject. Popular, inexpensive multi-element assay packages for reconnaissance prospecting are described. Since assaying involves knowledge of geochemical prospecting–rock, soil, and stream sampling–field procedures for obtaining these samples are explained.

In a world of micro-gold, pathfinder elements, and elemental associations, the gold pan and rock hammer are sensitive instruments when used correctly. These fundamental prospecting tools haven't changed much, but their application has. Beyond knocking off a chunk of rock and sending it in for an assay is the recognition of rock alteration and surface leaching as guides to mineralization. In a similar way, hunkering over a gold pan in a creek and concentrating only on finding gold flecks disregards the other eighty percent of the pan's information. The semi-heavies and heavy concentrates in the pan hold the secret to the upstream drainage. Careful analysis of these minerals may lead to gemstone pegmatites, diamond pipes, platinum metals, nickel, chromite, or base metals.

Included here are practical chapters on using a gold pan more effectively, recognizing rock alteration, and getting assays, plus descriptions of visits to important mining areas and advice from experts. One geologist-researcher tells how he uses creek names on topographic maps as guides to past mining activity. A Canadian prospector describes working on the successful search for diamonds in the Northwest Territories. In addition, I am including information about the valuable platinum-palladium metals and their geological settings, a reevaluation of the old rush for uranium and its resurgent market, and an examination of the secret activities behind a major new gold rush in Interior Alaska.

Well-known mining districts displaying a range of mineral deposit types are described. Circle, Alaska, illustrates the history and changing technology in placer mining for gold. Nevada's open pit gold mines show the recent attention to large-scale, disseminated deposits. Easily overlooked in the past, these modern bonanzas have subtle clues to their identification. Igneous rocks (crystallized magma) also continue to be favorable targets for finding ore bodies. The 1984 discovery of the Fort Knox granite-hosted gold deposit, in the middle of the eighty-year-old Fairbanks, Alaska Gold District, emphasizes the need to carefully examine all types and exposures of these rocks.

Exploration geologists, geology professors, miners, venture capitalists, drillers, and prospectors contributed their experiences and insights to these chapters. I've compiled the most up-to-date and useful information I feel is necessary for prospecting today. Even the final step of optioning and selling a prospect is covered for those who find a marketable mineral deposit. Prospecting can bring monetary rewards, but I know from experience that it takes hard work and a sensible plan.

This book does not attempt to cover rock and mineral identification or describe the full range of ore deposit types. Those subjects are best learned by taking prospecting or economic geology classes where they are available. It also helps to learn about the local geology of an area by joining a rock and mineral club and going on field trips. For even more information, rockhounding and gold prospecting books and magazines provide descriptions of specific areas, hints on what to look for, and suggestions on how to use what you find.

Ultimately, a mineral discovery comes from hiking around and getting to know the geology of an area. Prospecting is like a detective story, with clues and false leads, and the need for logical deduction, observation, and collection of facts. Analysis, research, and constant review are important. While the rare thrill of a major find is always an incentive, every day brings small, unexpected discoveries that lighten the pack and make the hike back to camp easier. Prospecting can be a life-long activity with many engrossing side areas to investigate. Fortunately, the quest can be as enjoyable as the prizes to be found.

1. How I Became a Prospector

Mosquitoes buzzed around us as my neighbor, Brian, and I clambered across a cool stream. We were prospecting in the heart of gold country near Fairbanks, Alaska. My friend carried a rifle, though we were only a half-mile from the road. Alders and willow grew thickly by the creek, but as we climbed out of the low area, a well-defined game trail led us through moss, labrador tea, and scrub spruce. A pile of rocks by a steep hillside and a squared claim post gray with age indicated that others had considered this a good prospect.

"When I was looking at early maps of this area, I noticed a small outcrop of granite was shown here." I wiped the sweat off my forehead and caught a mosquito in mid-air. "The latest maps overlooked it. It's a mystery that I wanted to investigate."

Brian took a swallow of water. "What's so important about granite?"

I hammered a chip off a nearby rock. "Geology books I've read say to check around granite for minerals, veins, and alteration. Besides, we're in the Fairbanks District where gold is related to two big domes of granite. One of those domes is named after Felix Pedro who first found gold in the creeks around it."

"You mean there's gold on Pedro Dome?"

"It's barren granite now," I explained. "Before it was eroded off, there were quartz veins and areas where the hot magma's contact with the schist caused minerals to form."

At that time, Fairbanks was a sleepy placer mining district, and the Fort Knox Gold Mine a few miles from Pedro Dome hadn't yet been discovered. This elephant among gold mines had been overlooked for almost a hundred years. Shear zones in the brittle granite rock had provided channels for gold-bearing fluids to follow. Once the cross-cutting quartz veins hidden under a blanket of moss were exposed, they stood out like lace doilies on a table.

When I had first read about mineral deposits, I noticed that igneous rocks were often mentioned as a primary target for exploration. On geological maps these rocks are colored red or pink and stand out from the pastel-colored sedimentary and metamorphic rocks. Once, when I visited an exploration geologist's office, I saw a map on the wall that had all the intrusives circled in red ink. The area within a mile of such rocks is considered prime exploration ground because extensions of the main body, known as cupolas, could be nearby, or buried but near the surface.

I originally learned about igneous rocks while hiking in the Sierra Nevada Mountains of California, where a huge area of granite (igneous rock composed of quartz and feldspar) forms the core of the range. Such large igneous bodies, called plutons or batholiths, are usually barren, but smaller bodies may be mineralized or have associated veins and mineral deposits in the surrounding rocks. Sometimes veins crop out, or hardening of the rocks indicate the hot effects of the magma when it was emplaced millions of years ago.

One of the great economic geologists, William Emmons, wrote a chapter on prospecting in his book, *Gold Deposits of the World*, published in 1937. He had seen a relationship between granites and gold deposits and he emphasized searching around these bodies for mineralization.

Nowadays, because of new discoveries in the Great Basin of the western United States and understanding of plate tectonic theory and its relation to mineral deposits, we have a greater range of favorable environments for gold deposits—not just igneous-related ones. However, the great variety of igneous rocks continues to be a major focus for mineral exploration.

"This isn't so bad," Brian observed. "Just pack a picnic lunch and bring enough mosquito repellent for an army. That way the bears won't eat you."

I had asked my friend Brian to come along partly because of his experience with a rifle. I wasn't yet comfortable out in the woods on my own.

This area had seen a lot of prospecting. Old miners' cabins with nearby piles of gravels that had been brought up from one hundred feet deep were hidden in the trees along the trail. Blanketing the lower hills and valley bottoms of the Fairbanks area, under a surface covering of vegetation, is a thick layer of wind-blown loess (silt) and debris from the Pleistocene era that is frozen as hard as cement by permafrost. To reach the ancient, buried creeks and the gold they held, the old-timers had to thaw the ground using wood fires or steam, sometimes to depths of one hundred feet or more.

Panning the present-day creeks which are on top of this barren silt yields only shiny mica from the schist. My plan was to examine the hillsides where outcrops gave clues to mineralization, and avoid the back-breaking work of digging in the valleys. I've learned that each area demands a different approach to prospecting. Figuring out how to prospect an area is the key to success.

"That gold is going to erode out of this hillside unless we get going," I joked. This certainly wasn't going to be easy, I thought. "Somewhere up this steep hillside is the source for these rocks," I reminded my friend.

We began to follow the granite "float"–rocks speckled with light and dark minerals that had come down from somewhere up the hillside. Birch and aspen rustled in the afternoon breeze. We finally reached an area where larger boulders cropped out. Above this, the slope became flatter and no more of the large rocks were evident. While Brian rested, I chipped off pieces of the boulders to examine. I noticed red streaks in the granite that could be cinnabar, and I knew gold often occurred with it.

"I think we've found the source," I declared. "We need to stake out this hillside and the creek below in case minerals eroded out and formed a downstream placer." I envisioned the miles of rocks that had been eroded off this area and the natural concentration of the heavier minerals in the valley below.

Granite altered by mineralizing fluids is crumbly and it easily breaks down to form clays. Many of the world-class copper deposits of the western U.S., such as those at Butte, Montana and Bisbee, Arizona, are related to altered granites. Prospectors were attracted first to the surface veins of silver and gold, but as they worked their way deeper into the altered rock, they found copper, and at the core was the barren granite.

"Should I say something like 'Eureka'?" Brian asked. We looked out over the birch forest and a spruce-covered valley below. I could already see the mine buildings and a tunnel leading into the hillside. We grinned at each other–what luck to hike a half-mile and strike it! The old-timers had somehow been all over this hillside and hadn't discovered this colorful granite,...or had they known something we didn't?

That was over fifteen years ago, and the birch trees still stand guard over that peaceful hillside above the valley. It took years for me to find out why the old-timers had walked away from this prospect. In the process I learned about mineral identification and ore deposits by taking numerous prospecting courses at the Mining Extension program of the University of Alaska in Fairbanks.

When I staked those first mining claims, I inaccurately described the township, range, section, and quarter section on the claim forms. Writing the legal description correctly took practice. With experience, I eventually mastered the fine points of where the northeast quarter of the southwest quarter was located, and which meridian my claims were in.

I got assay laboratory addresses and prices, and after selecting one of the analytical processes, I sent rock samples away for assays. I learned the chemical symbols for tungsten (W), mercury (Hg), silver (Ag), and gold (Au).

In the evening classes I attended, I met kindred spirits who were experienced placer miners, weekend recreational gold panners, and serious prospectors. There was Terry who had a golden touch for finding placer creeks, and losing them to someone else. Pat came from a gold rush family with claims on Deadwood and Switch Creeks. Roy had a passion for front-end loaders and CATs (bulldozers), and mining out on Preacher Creek.

Standing in front of his motley class, Professor James Madonna–Jim to us–lectured, cajoled, joked, and charmed us into his world.

"Wake up, Roy!" he would thunder. "What's the ore of lead?"

Tired from a day of physical work, Roy would lift his head from the table and reply, "Galena."

Always in tune with our lack of knowledge, Jim translated the difficult concepts. The variety of granites were like a "mixed salad." Some had more plagioclase or olivine or were sprinkled with "pepper" in the form of biotite or hornblende. Others were rich in quartz and potassium feldspars, like a salad with light-colored croutons and a creamy ranch dressing.

He drilled us about the intrusive and extrusive forms of magma using a camel-humped diagram spread across the blackboard. Oceanic basalts were the extrusive form of deeper peridotites, dunites, and other magmas. Island arc volcanic andesites came from deeper diorite and granodiorite magmas. Continental rhyolite lavas had granite and quartz monzonite roots. By the end of his classes we could all recite Bowen's Reaction Series–a complex process of mineral differentiation within the cooling magma that led to the different compositions of igneous rocks.

"Roger," I hear him ask, "what's the ore of zinc?" And I confidently answer, "sphalerite."

We all looked forward to the coffee and tea break so we could munch on cookies and exchange prospecting ideas and experiences. Afterwards we worked in pairs testing minerals with chemicals and a propane torch. These classes also taught me how to make and read geology maps, carry out geophysical surveys, and test soils for minerals.

We each went our separate ways with the knowledge we acquired. For some, the placer testing and mining techniques added to their operations in the creeks. Others continued as recreational prospectors, panning creeks for fun and a few nuggets. I set off on a life-long quest to find lode sources, the most difficult prize in prospecting.

After my initial "discovery" on the hillside, I invited a few geologists to visit the claims. They came with rock hammers and a scintillometer. The equipment showed that the rocks were slightly radioactive. However, the experts left scratching their heads about just what kind of deposit it could be. I was learning that geological processes produce an amazing variety of deposit types, any one of which I could stumble across in my search.

I never gave up on that curious granite with the red streaks. Over the years I traced its trend and found new places where it appeared. At one location it was cut by a small creek, and after getting the permits, I had a bulldozer come in and trench into the hillside to expose the spot. I remember sitting under a shady birch tree one summer looking over the hard rock that had been exposed. I had to admit I was no closer to understanding the mineralogy. There were smoky quartz veins and pyrites, and tantalizing areas of small black crystals. But no gold. Here I was in a famous gold district and there wasn't any of the typical colors—canary yellow from weathering antimony, or greens from oxidized arsenic—usually found with gold. I wondered why.

The state geologist was the first to suggest a solution to the puzzle. "You have a higher than normal tantalum content," he pointed out. "Next assay ask for a niobium analysis. It's often paired with tantalum."

When I sent in samples for assays, I usually asked for the "Gold Plus Thirty-six Elements" analysis, which lists the trace amounts of a broad range of elements. I paid a little extra for niobium in the next set of assays and, sure enough, there was a lot of niobium! It turned out that this igneous body was older than the other gold-producing Fairbanks granites.

The state geologist who had suggested checking for niobium was as curious about my granite as I was, and he sent in samples to be dated and analyzed. He got a whole-rock analysis that showed its composition was poor in quartz but high in thorium, zirconium, and rare earths. It had formed before the main gold-producing granites of the Fairbanks District. Although I had found an unusual new granite, it wasn't very rich in important minerals. However, it did get me started as a prospector by helping me learn about staking claims and reading assays.

SUMMARY

From this experience I learned a lot and I found professionals who could help. Taking prospecting classes taught me mineral identification, geology, and how deposits form. I learned which minerals tend to occur together. Gold usually has arsenic as a pathfinder element in the Fairbanks District. The earlier prospectors had recognized this and they had moved on. Now I prospect new areas more efficiently and I'm more receptive to the variety of minerals occurring in nature.

Prospecting has increased my enjoyment of the outdoors and appreciation of geological processes. The promise of a discovery leads me into new valleys. All I need for a hike is a light backpack containing these items: rock hammer, sample bags, gold pan and screens, mosquito repellent, compass, map, magnifying loupe, flares or a rifle (in case of bears or for signaling), rain gear, a snack, and water. That birch-covered hillside near Fairbanks, where I first began my education in prospecting, has led to years of healthy outdoor recreation.

2. Deposits: How Did They Get There?

A story about a hunting party composed of a geologist, an engineer, and a promoter pokes fun at geologists' preoccupation with the settings and processes behind ore deposits. The group finds a bear track. While the engineer sets out to shoot the bear, and the promoter goes to town to sell the hide, the geologist retraces the tracks to find out where the bear came from. Even with all that is already known about ore deposits, the conditions that created them are still of intense interest to geologists. Likewise, anyone who is interested in finding mineral deposits needs some understanding of basic geology and the theories about how ore deposits form.

Economic geology, the study of ore deposits and their valuable products, is constantly being refined as competing theories are tested by new discoveries. Knowing different deposit types gives prospecting a scientific basis and helps prospectors target higher-potential areas for exploration. By visiting and studying known mining districts, the prospector and geologist can examine the complex factors underlying mineral deposition. Deposits come to life by taking field trips or following road-log guidebooks, and by having direct contact with the rocks. Samples collected from mine dumps and roadcuts can later be studied and compared to written accounts about the deposit setting and origin.

CLASSIFICATION OF ORE DEPOSITS

Theories about how ore deposits originated and how deposits should be classified have been hotly debated for centuries. Now a broad consensus among geologists identifies general categories of mineral deposit origins and deposit "types" or "models" that emphasize the conditions under which the deposits formed. I remember when a friendly geologist first loaned me an economic geology textbook.

"The ideas are a little out of date," he warned me, "but it shows the basic types of deposits." As I learned about the scientific theories underlying the formation of ore deposits, I began to understand and appreciate these complex creations of the physical world.

Genetic models based on magmas, or melted crustal and mantle rocks, reached their peak in Waldemar Lindgren's influential *Mineral Deposits*, which was first published in 1913. Even then, sedimentary processes, chemical processes, and hydrothermal solutions from surface, ground, sea, and magmatic water, were recognized as important in ore deposition. Present-day active volcanic cones, sea floor spreading centers, and hot springs illustrate some forms of mineral deposition, and recently, many ore deposits have been explained in relation to plate tectonic settings.

MAGMA AND ORE DEPOSITS

That magma had a role in creating ore deposits was advocated by economic geologists beginning with James Hutton's *Theory of the Earth*, published in 1788. The heat engine of an intrusive magma causes a range of reactions within the magma at the contact zone, and at great distances from the body. Ore deposits that occur within the intrusive body include porphyry copper, molybdenum, gemstone pegmatites, tin veins, and gold.

Prospectors and geologists love finding undiscovered granites (or granodiorites, quartz monzonites, or diorites) because of the possibility of a wide variety of associated deposits. Even today, these "salt-and-pepper" rocks continue to be the most compelling target for prospecting. My early prospecting was essentially an investigation of all the known igneous outcrops in the Fairbanks Mining District (see chapter 5, "Fort Knox, The One That Got Away").

The best-known deposits occurring within intrusives are the porphyry coppers that contain disseminated copper, usually in the form of chalcopyrites, and many of these deposits extend into the surrounding schist or sedimentary rocks. The early copper discovered in the southwestern United States was in granite that was characterized by phenocrysts (mineral grains with crystal form) in a very fine-grained groundmass—a distinctive rock texture called "porphyry" by geologists. Composition of the host magma varied from diorite to granodiorite to quartz monzonite to granite, with most being fairly quartz-rich igneous rocks.

As higher grade native copper deposits in Michigan's Keweenaw Peninsula were mined out, the low grade copper porphyries became more economic to mine and continue to be a major source of copper. Typical examples of such deposits are in areas where oceanic plates were subducted under continental or island arc margins, creating intrusions in the upper mantle, as in the southwestern United States and the Andean Cordillera.

Alteration features within and around porphyry coppers are like the envelopes around a candle flame (see chapter 16, "Rock and Mineral Alteration"). The primary zone may encompass an area from 1,000 feet to 9,000 feet across with widespread secondary effects coloring the surrounding rocks and producing an extensive geochemical halo. The deposits have a distinctive geophysical signature because of the conductive copper minerals and creation of magnetite in the surrounding rocks. The mines at Butte, Montana, Bingham Canyon, Utah, and Ruth (Ely), Nevada, have viewing platforms that overlook their open pits. At Bisbee, Arizona, a surface tour takes visitors to the edge of the open pit to look at the red, orange, white, and yellow colors–evidence of the oxidation of iron minerals and altered feldspar clays.

Recognizing altered feldspars, white mica or sericite, and "ragged" biotite in these igneous rocks, is an important clue to identifying potentially productive areas to prospect. For a class assignment, my ore deposits professor handed out trays of igneous rocks from porphyry copper deposits at Yerington, Nevada, and Bingham Canyon, Utah. We had to group the rocks from the barren core of the granite (unaltered feldspars that are hard and fresh), through the mineralized intermediate zone of clay-altered feldspars and ragged biotite with chalcopyrite in minute fractures, to the pyritic and outer barren zones.

Bins at the Bingham Canyon Museum and Overlook furnish visitors with pieces of the barren pyrites that are ubiquitous in these deposits. However, locating examples of granite porphyry with chalcopyrites is rather difficult, because this rock is sent to the mill. Some examples may be found along the eighteen-mile railroad right-of-way that runs from the Ruth/Robinson Copper District near Ely, Nevada to the reclaimed smelter area at McGill. Otherwise, rock shops that carry copper minerals from these deposits may have inexpensive specimens in matrix that illustrate the granitic porphyry.

PEGMATITES: CRYSTAL POCKETS FROM AN ENRICHED GRANITE MELT

Pegmatites are small and irregular deposits of crystals that formed as residual liquids from a magmatic melt cooled, and are found within and near magma that crystallized deep under the surface (see chapter 20, "Black Hills Pegmatites"). These highly sought after gemstone and crystal prizes may contain beryls, lithium minerals such as tourmaline and lepidolite, topaz, rose quartz, or other rare minerals. Since their deposition occurs at great depth, the surrounding rocks are usually highly metamorphosed and brittle. These rocks fracture and provide pockets for the liquids that create the pegmatites. When exposed by erosion, these resistant outcrops of light-colored feldspars, quartz, and muscovite mica advertise their pegmatite treasures.

One of the distinctive pegmatite features is the large crystal size, which probably results from growing in a water-rich magma. Creamy-white feldspars intergrown with quartz (graphic granite) and abundant muscovite mica indicate their late, watery composition. "Fertile" granites enriched in fluorine, boron, lithium, beryllium, and niobium are good targets to explore. Pegmatites occur at various distances from their granitic source, and differentiate into simpler compositions as they travel farther from their origin. So unzoned pegmatites, containing only quartz, feldspar, and mica, may provide clues in the search for more complex zoned ones that have larger crystals in their intermediate zones and quartz cores.

Located in a pegmatite belt in northern New Mexico, the famous Harding Pegmatite near Taos, owned by the University of New Mexico, provides visitors with an outstanding example of pegmatite zoning and mineralogy. This mineralized area is 2,500 feet long and up to 500 feet wide and contains more than fifty pegmatite minerals, including niobium and tantalum, rose muscovite, lepidolite, beryl, and sodium feldspars. Visitors can obtain information and permission to collect specimens by writing to:

EARTH AND PLANETARY SCIENCES
University of New Mexico/Northrop Hall
200 Yale Boulevard
Albuquerque, New Mexico 87131-1116
(505) 277-4204 / (505) 277-8843 Fax
http://epswww.unm.edu

TIN GRANITES

Tin granites may generate tin pegmatites as well as a variety of tin deposits in the form of veins, in muscovite- and quartz-rich apical areas of the granite called greisen, and in placers eroded from these lode sources. Many tin deposits are found because tin oxide, cassiterite, is concentrated as gray pebbles along with gold in stream placers.

While tin granites tend to have higher silica (averaging around seventy-three percent), and elevated concentrations of tin (Russians found that sixteen to thirty parts per million was anomalous), lithium, fluorine, and chlorine are better indicators. These granites may also be enriched with boron, niobium, tantalum, uranium, thorium, and rare earth elements.

Cornwall in southwest England supplied Mediterranean Bronze Age metallurgists with tin. The region had veins within the extensive granite, in greisen and stockwork veins at the cupolas of the batholith, and in deposits at the contact zone and in the surrounding rocks. Mineral zoning in the Cornwall deposits placed tin, with or without tungsten, close to the granite, copper, and lead-zinc deposits further away. Little remains of this famous mining area except stone and brick smelter chimneys, but the granite batholith still can be examined in many exposures.

DEPOSITS WITHIN THE MAGMA CHAMBER

At the other end of the igneous spectrum of ore deposits are the nickel, chromium, titanium, magnetite and platinum group metals found in magmas derived from mantle sources called mafic and ultramafic magma because of their high magnesium and iron (ferric) and low silica content. These "dark" rocks are composed of olivine, pyroxene, plagioclase, and hornblende and are depleted in quartz (see chapter 21, "Dark Rocks and Platinum Metals"). During the cooling of the magma chamber, the minerals crystallize from the melt into layers or segregations, so they are called "layered magmatic" deposits.

The International Nickel Company (Inco) spent seven years searching for nickel in peridotite, an olivine-rich igneous rock, in Manitoba, Canada, before finding the deposit that became the Thompson Mine in 1956. The principal ore of nickel, pentlandite, a nickel-iron-sulfide, is found in small quantities in peridotite and serpentinite (altered peridotite) and related rocks along with pyrrhotite and chalcopyrite. The unique Sudbury area of Ontario, Canada was the result of a meteorite impact melting the crust and forming nickel-copper deposits around its oval margin. This was Inco's major nickel source until the Thompson Mine discovery showed that the small amounts of nickel found in peridotites could be remobilized during metamorphism into richer concentrations in nearby pyritic black shales.

Another kind of layered magmatic deposit happens when chromitite (rock containing chromite) layers form in dunite, an ultramafic rock with a very high olivine content. These rocks weather to a dun color while the black chromite masses and disseminations are resistant and stand out. Stream pan concentrates will turn up the heavy chromite grains.

Chromium derived from chromite is used to strengthen stainless steels, for chrome plating, and in a variety of other alloys with iron, nickel, and cobalt. South Africa's Bushveld Complex contains three-quarters of the world's chromium and Zimbabwe has most of the remaining higher grade deposits in its Great Dyke. The Stillwater Complex in Montana, a layered ultramafic belt twenty-eight miles long, contains about eighty percent of the U.S. chromite reserves. Since other countries mine chromium more economically, this deposit remains undeveloped.

Palladium and platinum are also found in the Stillwater Complex, and the Stillwater Mine currently supplies about ten percent of the U.S.'s needs. As demand has increased for these metals, which are used in automotive catalytic converters to clean up emissions, they have become more valuable and attractive for exploration. Sparse disseminations of platinum with chromite in dunites create platinum placer deposits while sulfide-rich occurrences in ultramafic magma chambers make better lodes. The Stillwater layered intrusion contains disseminated, platinum-bearing sulfides as a narrow zone halfway above the base of the body. Fortunately, the body is tipped on its side, making the valuable zone accessible. This area near Nye, Montana has Forest Service roads that provide access to the ultramafic rocks in the Beartooth Mountains.

DEPOSITS OUTSIDE THE MAGMA CHAMBER

Skarn, another kind of deposit related to emplacement of a hot magma, forms at the edge of an intrusive in the contact zone and reactive limestone beds (see chapter 18, "Skarn: Where the Magma Hits the Marble"). Also termed tactite, pyrometasomatic, and contact metamorphic, skarn is a replacement deposit formed at high temperatures in carbonate-rich sediments. Geologists carry weak hydrochloric acid for squirting on rocks to test their carbonate content, and finding such reactive bedding dipping toward an intrusive raises the potential for a skarn.

These deposits can contain a variety of metals, including tungsten, copper, molybdenum, lead-zinc-silver, iron, or gold, either singly or in combinations. Molybdenum normally is present closest to the magma, while tungsten, lead-zinc, gold, and iron are deposited at greater distances and cooler temperatures. These irregularly shaped and erratically distributed deposits may be high-grade targets, and are often indicated by garnet and epidote that color the area dull red and green.

Inactive tungsten mines west and northwest of Bishop, California contain examples of a tungsten skarn containing dark-colored calc-silicates (garnet, epidote, hornblende, and pyroxene) formed by magma contacting impure limestone and dolomite, sprinkled with scheelite, the ore of tungsten, which can be seen under an ultraviolet lamp. Because tungsten's specific gravity is around 6.0, just a little lighter than lead, (gold's is around 19), scheelite is heavy enough to be found in placer concentrates.

HYDROTHERMAL SOLUTIONS AND DEPTH CLASSIFICATIONS

Farther from the magma heat source, sometimes miles, hot circulating aqueous solutions called hydrothermal fluids may chemically replace carbonate rocks with minerals. The lead-silver ores of Leadville, Colorado and Tombstone, Arizona are examples of this process. Structurally favorable limestones and dolomites, heated fluids from the magma, permeable openings from earlier replacement and fracturing, and oxidation and remobilization of minerals, were all factors in creating the deposits.

In the early 1900s, the geologist Waldemar Lindgren developed a depth classification of mineral deposits that is still in use. He defined hypothermal deposits as those formed at depth under high temperatures and pressures, mesothermal deposits as those formed at intermediate temperatures and pressures, and epithermal or near-surface deposits as those found in the hot springs environment and low pressures of faults and rock openings. In the range of deposits discussed in his book, *Mineral Deposits*, chromite and platinum metals would have been formed at the greatest depths in ultramafic magma chambers, whereas most skarn, replacement, and precious metal quartz veins would have been mesothermal.

The author collects examples of rock alteration and mineralization from a mine dump in Tonopah, Nevada. Historic mining districts provide opportunities for learning about the geology surrounding ore deposits.

Most interesting to Lindgren and other western U.S. geologists were shallow epithermal bonanza gold and silver deposits. Based on a hot springs model, circulating thermal waters leached metals from surrounding strata and redeposited them, as the temperatures and pressures decreased, near the surface (see chapter 17, "Yellowstone's Geysers and Hot Springs"). Carlin (Nevada)-type disseminated gold deposits in reactive limestone and dolomite beds had a similar origin, coming up through faults and channelways created as the Basin and Range was being pulled apart or rifting (see chapter 9, "Nevada Monsters: Carlin's Micro-Gold").

I frequently visit the Comstock mines around Virginia City, Nevada, to examine the alteration and mineralization style of an epithermal deposit. In the 1870s and 1880s, the still-active hydrothermal system underground made the mines so hot and sulfurous that miners could only work for a few minutes at a time. These days, examples of the now cool cross-cutting quartz veins glittering with crystals and gray sulfides still litter the inactive open pits and hillsides.

The state park at the preserved ghost town of Bodie, California, has regular mine tours of the near-surface quartz-veined areas on Standard Hill. These "stockwork" veins are a closely-spaced network of veinlets created by a shallow magma. Similarly, the silver veins of Tonopah and the gold of Goldfield, Nevada, are related to shallow calderas, another variation of the epithermal deposit. These towns have many old mine dumps showing typical clay alteration and quartz-filled rock openings.

UNDERSEA BLACK SMOKERS OR VOLCANOGENIC MASSIVE SULFIDE DEPOSITS

Base metal (copper, lead, and zinc) deposits related to rifting sea-floor centers share similarities with the hot springs hydrothermal model. Copper-zinc and zinc-lead-copper volcanogenic massive sulfide (VMS) deposits are actively forming on the sea floor. Characteristic of this setting are basalts or pillow basalts that may be later metamorphosed into greenstones. Felsic (feldspar and silica-rich) magma is commonly believed to have been a heat source that circulated the sea water through fractures in the basalts.

First recognized and named "Koroko" and "Besshi" from Japanese mines–terms still used by some–this model has led to many famous base metal mines being reinterpreted as VMS deposits. The best examples in the United States are at Jerome, Arizona, at the United Verde and United Verde Extension (or Daisy) Mines. The state museum in Jerome has a geologic model showing the deposit formation, and specimens of the important rocks can be gathered from nearby roadcuts and waste dumps.

LOW TEMPERATURE SEDIMENTARY DEPOSITS

The circulation of low temperature fluids (sometimes less than a 212°F or 100° C) created many ore deposits, among them the lead-zinc ores of the Mississippi Valley, the Colorado Plateau sandstone-hosted uranium, and the Lake Superior Copper District copper in Michigan. Circulating brines and percolating ground-waters, followed by precipitation of minerals because of the reducing environment and other factors, led to these stratabound, flat-lying deposits. Sedimentary layers may not always look like promising places for mineral deposits, but in the right conditions, they may contain easily-mined concentrations of lead, zinc, uranium, or copper.

Carbonate strata that were once algal reefs were important deposition sites in the Mississippi Valley Tri-State District of Missouri, Oklahoma, and Kansas where sphalerite (zinc) and galena (lead) deposits were discovered in 1720. Geologists believe that prolonged movement of sulfate-rich solutions leached trace metals from sediments and concentrated them in the porous limestones.

Other stratigraphically controlled deposits are found in the Colorado Plateau and Four Corners area of the southwestern United States, where groundwater leached uranium from distant granitic source rocks and redeposited it in sandstone basins (see chapter 23, "Uranium, a Second Look"). Michigan's Keweenaw Peninsula copper deposits, which formed in shallow continental or marine basins, may also be related to leaching of trace copper by saline groundwater.

METAMORPHIC GOLD-QUARTZ IN FAULTS AND SHEAR ZONES

Having dismissed a lot of "bull" quartz—a miner's term for a blocky and barren kind of quartz—I wasn't prepared for the fault-related masses of gold-bearing quartz of the California Mother Lode. Also seen at Yellowknife (in the Northwest Territories of Canada) and Juneau, Alaska, such "veins" are the result of metamorphic and hydrothermal processes that squeeze solutions containing silica, gold, and sulfides into any available openings. These unlikely-looking quartz masses may also have chlorite, white mica, pyrite, arsenopyrite, and carbonate in the form of ankerite. Now, even if the quartz is not vein quartz, I know to examine it more carefully for indicators such as chlorite or arsenopyrites.

On a visit to the Royal Mountain King Mine near Copperopolis in the Mother Lode, I collected specimens of "gray ore" that had formed when slate along a fracture zone had been altered to a gray mass of irregular chunks of ankerite and blocky quartz. In addition, small cubes of pyrite and arsenopyrite were sprinkled throughout the ore. Similar examples of the brittle slate host rock, with its quartz and carbonates, may be found in some of the inactive mine dumps and roadcuts along California State Highway 49.

The quartz of the California Mother Lode was formed by metamorphic pressures and hydrothermal fluids. This mine dump near Bear Valley, California has examples of this blocky quartz.

OTHER DEPOSITS

Other deposit types not included in this brief overview include titanium, iron formation gold (such as at the Homestake Mine at Lead, South Dakota), niobium-tantalum carbonatites, base metal veins (Butte, Montana, has good examples), mercury (California's New Almaden Mine State Park south of San Jose preserves many old mercury mines), cobalt veins, diamondiferous kimberlite pipes, and copper shale. There are also many other deposits, which economic geology texts illustrate.

PLATE TECTONICS AND ORE DEPOSITS

New concepts of how mineral deposits are related to tectonic settings now provide another method for focusing regional exploration. Mineralization is related to mantle plume hot spot activity (Hawaii and

Yellowstone National Park are modern examples), rifting (in extensional areas such as the Basin and Range of the western U.S., as well as sea-floor spreading centers), shallow subduction creating island arc volcanics and their back-arc basins, and deeper subduction under continental margins (porphyry copper and molybdenum deposits).

The earth's crust is constantly in motion, recycling material from the ocean as it subducts under or obducts onto the continents. When oceanic plates are being subducted under continents, they create magma, which, because of its low density relative to the surrounding rocks, rises upward like giant balloons incorporating material and cooling to form igneous rocks, or to reach the surface as volcanos and extrusive rocks. A continent may be either made up of plates from far different areas that were patched together, or part of a land mass that was rifted apart. Failed rifts are weak areas in the earth's crust that provide sites for hydrothermal fluid circulation or emplacement of magma. By recognizing a particular tectonic setting, a prospector can visualize the potential deposits to be found in those rock types.

PLACER AND PALEOPLACER DEPOSITS

Placers and paleoplacers created by mechanical concentration are another important deposit type in which resistant and heavy minerals are eroded from their original source into streams, deltas, and beaches. Gold placers were the first deposits worked by people, and these continue to be attractive for small miners (see chapter 7, "Before the Klondike: Circle, Alaska"). Alluvial gold placers in gulches, streams, and rivers are the richest, though Nome's marine beaches were also very productive. Cassiterite (tin), gemstones, platinum, and scheelite (tungsten) also form economic placer deposits.

Unglaciated, deeply weathered hills, and gradual uplift provide the best conditions for the formation of gold-bearing stream placers. Mechanical and chemical weathering processes working over long periods of time set the gold free from its source deposits or country rocks. In some places extensive glaciation destroyed the stream placers, except where there was little ground disturbance. A mature topography with rounded hills, rather than sharp mountains or flat terrain far from mountain systems, is more favorable for the creation of stream placer deposits.

STRUCTURES RELATED TO MINERAL DEPOSITS

Recognizing the importance of having favorable geologic structures–fissures, faults, shear zones, brecciated areas, impermeable barriers, folds, and anticlines–to impound and localize mineralized fluids is essential to understanding ore deposit formation. Such features are most easily observed in a mining area because of the exposed workings. The Comstock, Nevada, deposits, for instance, are strung out along a normal fault. In a similar way, the California Mother Lode gold-quartz veins are in deep faults in brittle slates and granites. On the other hand, the Bendigo, Australia gold-bearing quartz lenses were found in the crests of over ten stacked anticlines.

Tombstone, Arizona, illustrates multiple factors that may be involved in creating an ore body: impermeable anticlinal or folded beds were structural traps; faults and dikes formed conduits for mineralized fluids to travel through; and chemically reactive and permeable limestones were favorable replacement sites for the deposits. Later remobilization and secondary enrichment by circulating waters increased the grade of the ore by leaching out the sulfides and redepositing them as "horn silver," manganese oxide, argentite, chalcocite, chrysocolla, and other oxide minerals. In all, nine factors have been identified as influencing mineralization, not unusual for an ore deposit.

HISTORIC MINING DISTRICTS

On a trip to Arizona with my wife, I reluctantly stopped at Tombstone, dreading the boardwalks lined with curio shops, the noisy shoot-outs, and the daily stagecoach tours. But behind the tourist-dominated town, I found that the hills were pock-marked with rubble heaps from the old mines. There were limestones, silicified shales, and quartzites that readily fractured and would have promoted the migration of ore-bearing solutions.

I collected chunks of limestone heavy with sooty manganese, and other rocks that had colorfully-lined cavities–examples of leaching and secondary enrichment of silver. A granodiorite, once a hot magma that had baked and hardened the ground, cropped out in several places. When I later read about the deposit, I could envision the concentration of rich silver ores along the Tranquility Fault and the Contention and Empire Dikes, which are now scooped-out open pits.

Field trips to almost any area of the world provide examples of geologic processes, important structures, and tectonic environments. For instance, a visit to Yellowstone National Park and its geysers and hot springs illustrates the beginnings of a bonanza silver or gold deposit. Knowing the broad categories of mineral deposit models allows a prospector to focus attention on productive areas. The settings of famous mining districts illustrate deposit types and expose the unique geologic conditions underlying their formation–furnishing examples to study like so many books waiting to be read.

REFERENCES

A condensed list of deposit types is given in "Canadian Mineral Deposit Types: A Geological Synopsis", edited by O. R. Eckstrand (Geological Survey of Canada, *Economic Geology Report 36*, 1984).

Economic Mineral Deposits by Mead L. Jensen and Alan M. Bateman (John Wiley and Sons) is a textbook on the subject of economic geology. Many older books with economic geology titles are also good reading, though they do not incorporate tectonic settings in their discussions.

The Porphyry Coppers (1933) by A. B. Parsons, and his update, *The Porphyry Coppers in 1956*, described many North American examples. Spencer R. Titley and Carol L. Hicks edited *Geology of the Porphyry Copper Deposits Southwestern North America* (1966), another classic on the subject.

Prospecting for Gemstones and Minerals by John Sinkankas explains mineral deposits and focuses on collecting specimens, especially from pegmatites. Stephen Voynick wrote "The Harding Pegmatite Mine" in the February, 1997 issue of *Rock and Gem* magazine.

Pegmatites in Maine and other parts of New England and those near Custer in the southern Black Hills of South Dakota are best found by visiting local rock shops in those areas for information. Lincoln R. Page and others compiled "Pegmatite Investigations 1942-1945 Black Hills, South Dakota" (USGS *Professional Paper 247*); J. B. Hanley and others did similar work on Colorado, Wyoming, and Utah (USGS *Professional Paper 227*, 1950); and E. N. Cameron and others covered the New England pegmatites (USGS *Professional Paper 255*, 1954). Richard Jahns discussed pegmatite formation and composition in "The Study of Pegmatites" in *Economic Geology's Fiftieth Anniversary Volume*, edited by Alan Bateman (1955).

Roger G. Taylor reviewed all the tin deposit types in his book *Geology of Tin Deposits* (Elsevier Scientific Publishing Company, 1979).

F.J. Sawkins' *Mineral Deposits in Relation to Plate Tectonics* (Springer-Verlag, 1990) explains the new discoveries in this field. *Economic Deposits and Their Tectonic Setting* by Charles S. Hutchison (1983) has many diagrams illustrating these processes.

Robert W. Boyle thoroughly discusses placers in "The Geochemistry of Gold and Its Deposits" (Geological Survey of Canada *Bulletin 280*, 1979).

Ore Deposits as Related to Structural Features (Princeton University Press, NJ, 1942), edited by W.H. Newhouse, documents how structure localized ore deposits in many mining districts.

USGS *Bulletin 2042-B*, "The Bisbee Group of the Tombstone Hills, Southeastern Arizona–Stratigraphy, Structure, Metamorphism, and Mineralization" by Eric R. Force (1996), explains the conditions that formed the Tombstone deposits.

3. Discovery Stories

I led the two geologists, Dave and Bill, along a rubble slope high above a forested valley, watching the slippery caribou moss and talus with each step. Dark siltstones that had not quite metamorphosed into slate clattered downslope. I reached a mossy island and paused to hammer on rocks, selecting out ones showing changes in color, unusual fractures, and veining.

"Somewhere around here," I assured them, "there were veins in the siltstone that assayed silver and gold."

Bill sat down contentedly and began packing a white muslin bag full of the nearest rocks. He was happy just to be out of the office and on the hunt. It was a sunny day with puffy clouds in the distance and we had a million-dollar view of the Tolovana River Valley, sixty miles north of Fairbanks, Alaska. Dave coursed out ahead, a lean and muscular picture of health and vitality. He carried a heavy crack hammer that sent rock fragments flying with each blow.

"Over here!" Dave called. We scrambled to where thin white veins appeared in the dark rocks. Bill wrote an identifying number on the sample bag and sat down amid the rubble. "Granite begins just through these spruce," Dave informed us. A jumbled boulder pile indicated the changed rocks. I'd been through there before and hadn't found much of interest, yet Bill and Dave seemed attracted to the barren pile of igneous rocks. They hammered incessantly, hardly moving from one area. I moved on into a birch grove where more granite held some soil for the trees to grow on.

"Sulfides," Dave announced. "Looks like arsenopyrites." He held out a ragged vein colored with greens and blues. "Probably where those veins in the siltstone came from." I looked at the sample in disbelief. Hadn't I been over that area and searched those same rocks?

"You really have to spend time cracking rocks when you get indicators like those veins in the siltstone," Bill advised me.

Now that I knew what to look for I soon found my own rich sections of the vein. These would assay seven percent silver with minor gold, but since the vein was only two inches wide it probably wouldn't become a mine. Perhaps deeper in the mountain the vein became richer. Finding that out would take me far beyond the discovery and into the world of mining companies, drillers, and world metal markets.

The small amount of gold in the vein had contributed to a gold placer deposit that filled the valley below and was under my claims. Far down the valley I could see the old buildings and gravel workings that had first given me a clue that there was a lode source upstream. The placer miners just hadn't followed up the trail.

Dave, Bill, and I shouldered our heavy packs and picked our way carefully back up the mountainside. Fortunately, tired muscles and a heavy pack suddenly become insignificant after making a discovery. I call it a "leap of the heart." After months and sometimes years of searching, finally tracking down a hidden mineral deposit is wonderfully satisfying and makes the detective work and perseverance worthwhile. And anyone can join such a search. The basic tools are a rock hammer, magnifying lens, and knowledge of minerals, but it also helps to have a curious mind, willingness to hike, and a love of the outdoors.

GEORGE CARMACK

In *Carmack of the Yukon* (Epicenter Press, 1990) James Albert Johnson relates how George Carmack packed other peoples' supplies over the Chilcoot Pass for years in order to get enough of a grubstake to take off on his own. He lived with the Athabascan Indians and worked with Skookum Jim and Tagish Charlie as a packer. He taught them how to pan for gold, and to look for bedrock in the creek where natural riffles would trap gold. The eleven years he spent searching for a rich gold placer ended when they took a break from logging on the Klondike River to prospect its tributaries. After days of panning flour gold and grain-sized pieces on what became Bonanza Creek, they found nuggets in the exposed bedrock.

The odds of finding a mineral deposit are said to be one thousand to one or greater. Carmack had checked hundreds of streams and worked marginal claims, always believing that with hard work and a thorough checking of each creek he would find his bonanza. Whenever I watch geologists like Bill and Dave at work, I am reminded of the importance of exercising patience and care with every prospect.

TOM WALSH

Tom Walsh and his discovery of the Camp Bird Mine near Ouray, Colorado is another example of a prospector who learned his lessons and made the odds favor him. His daughter, Evalyn Walsh McLean, wrote about his discovery in *Father Struck It Rich* (Little, Brown and Co., 1936). While working as a carpenter in Deadwood, South Dakota during the Homestake boom in the 1870s, Walsh befriended "Smoky" Jones, a prospector who later tried to interest him in a mine. Walsh's mining expert friends discouraged him from the investment, so he passed up a share in the fabulously rich Homestake Mine.

From the Black Hills he returned to his home state of Colorado where Leadville was booming. He studied minerals and went prospecting. On a trip to the Frying Pan District he rested in an abandoned cabin near a barren prospect. The miners had tunneled into granite, unaware that they had built their cabin directly on a rich quartz vein. Walsh noticed the quartz and found its origin. That claim gave him his first taste of success, and even though he moved to Denver, married, and invested in real estate, prospecting was in his blood.

A few years later the family moved to Ouray, in southwestern Colorado, and Walsh took his young daughter to look over a claim high in the mountains. Snow covered the claim and tunnel, so after they returned home, he sent a friend up to sample it. Walsh was sick at the time and couldn't ride the nine miles to the prospect. But after his friend brought back barren-looking samples, Walsh wasn't satisfied and he forced himself to go back up into the mountains. Even though Ouray was a silver camp, Walsh recognized gold tellurides in the vein of quartz from the mountainside claim, and he noticed that the waste material outside the tunnel had similar high-grade ore on it. His experience recognizing minerals, and his insistence on doing his own sampling led him to the incredibly rich Camp Bird Mine.

BOB WOMACK

Money Mountain by Marshall Sprague (Little, Brown and Co., 1953) tells the story of how Bob Womack began tracing gold-bearing float up the creeks in the high cattle country of Cripple Creek, Colorado, in the 1870s. He knew there was placer gold in the creeks but he wanted to find the source. He slowly searched up Poverty Gulch for two miles. People ridiculed the hard-drinking Womack and his stories about a bonanza.

After eight years of prospecting, he staked a claim. A dentist in Colorado Springs grubstaked him after looking into the geology of the area and finding that it was an ancient caldera, or collapsed magma chamber. Working underground near his first claim, Womack finally traced the float to a hidden outcrop. No one showed any interest in the dull gray rock, so he left it in a grain store window in Colorado Springs. Finally, two men with mineral identification experience recognized the gold-silver tellurides in the rocks, and by 1891 the Cripple Creek Mines came into production. Bob Womack's persistence over the years made the discovery possible.

JACK SMITH

In 1900, Jack "the Arizona Centipede" Smith and Clarence Warner were prospecting the upper Chitina River area in Alaska for copper, and the going was getting rough, when Smith noticed an unusual green patch high on the mountain. "You can go look at that piece of goat pasture," his companion asserted, "but I'm not moving from here." His ankle was hurting and he wanted to return to camp.

"Look at this silvery float in the creek!" Smith exclaimed. That got Warner interested and they followed it up the creek where it led to the goat pasture. Copper ore outcropped in pinnacles sixty-five feet high along the peak of a sharp ridge, and the talus slopes held accumulations of more rich copper. The bonanza Kennecott Copper Mine was discovered because Smith had insisted on climbing up to investigate a green patch that turned out to be green malachite that covered almost pure chalcocite, a source of copper (*The Copper Spike* by Lone Jensen, 1975) For more information, write to HaHa, Inc., 3605 Arctic Boulevard #1431, Anchorage, Alaska 99503, (907) 243-1708, or visit http://www.alaskagold.com/haha/.

SHERMAN WILHELM

Sherman Wilhelm built a camp wagon for his family of seven kids and wife so they could travel together in the western states as he prospected in the early 1900s. Walt Wilhelm, his son, told the story in *Last Rig to Battle Mountain* (William Morrow and Company, 1970). For thirteen years the family moved from Colorado to California, Oregon, and then Nevada, living a subsistence lifestyle that kept the family together and happy.

In the summer of 1909 Wilhelm had a company of fifteen men working south of Battle Mountain, Nevada on the Betty O'Neal Lead-Silver Mine. They weren't making a good enough return, however, so he shut the operation down. A few years later another man found the mine's hidden ore body and turned it into a good mine.

Meanwhile, Wilhelm had bought two claims a few miles to the west, in Philadelphia Canyon near Galena Creek. There were small veins everywhere, and he felt a tunnel would strike good ore. Before commencing development work, he went over the ground, staking additional claims and looking at the geology.

One hot day in August while resting, he noticed reddish quartz sticking up under a sage brush. When he kicked it, the rock bent. He pulled up the sage, kicked the rock again, and it bent the other way. Only gold, being malleable, could do this. He had a claim form in his pocket and he put in claims to cover the area, naming it the Cussin' Ken Claim after one of his sons. Quoted in the Reese River Reveille, Wilhelm expressed his feelings after finding his long sought after wealth: "...I have worked for many years to get fixed so I wouldn't have to work, [and] now we have a property so rich I am compelled to work it personally on account of its richness."

He and his company worked the claims for several years before he sold his interest and moved his family to Southern California. Since then, Battle Mountain has produced many famous copper and gold mines from the Copper Basin and Copper Canyon areas where Wilhelm found his fortune.

HISTORIC MINING AREAS AND ABANDONED CLAIMS

Finding an ore deposit today is usually preceded by research into the mining history of an area for ideas on the types of deposits to expect. Most of the obvious old mines have been checked, but there are still ghost towns that haven't been thoroughly reevaluated. These abandoned mining areas should be investigated in ever-widening circles, since faults and mineralized trends often are displaced or may extend for miles beyond their obvious outcrop, and a lode source on one side of a hill may extend to the other side as well. For example, watchmen helping on the Kennecott Mine prospected the surrounding mountainside and found an extension of the deposit. George Hearst bought all the claims surrounding his Homestake Mine in Lead, South Dakota, but he only later found out that the main deposit had many rich extensions under these barren looking claims which didn't outcrop on the surface.

Abandoned claims near the Trans-Canada Highway north of Lake Superior got the attention of two prospectors researching claim status. The area had been trenched and drilled but gold values were erratic. The Canadian prospectors met and became partners while each was restaking the claims during a cold December in 1979. By 1985, the mine began the first of twenty years of production (*Golden Giant: Hemlo and the Rush for Canada's Gold* by Matthew Hart, 1985). In this case the prospectors recognized the potential of the ground and they acted when it came open.

MY EXPERIENCE

I'm working on an area in Interior Alaska that has a long history of gold placers—in fact many claims are still active and a small trommel wash plant continues to rework the gravels. A massive federal study was done on the area before part of it was designated a national recreation area where mining was prohibited. I hired a Super Cub pilot to fly me over the active mining lands to check on access and to map the placer workings.

"There's a cabin down there," the pilot yelled over the noisy engine. He circled around so I could spot the abandoned cabin in the spruce. I marked its location on a map, and noted where roads and trails entered the area. "I used to come up that road to hunt caribou," he told me, pointing to a rocky track leading up the ridge. "Not too many of the caribou are left," he commented, "but plenty of bears still roam these hills."

Later in the summer when the snow patches had melted and wildflowers were out in profusion, I drove my old Jeep Wagoneer up to the ridgetop and camped out. Recreational gold miners came by on their four-wheelers.

"We're panning the creeks," they told me. "A lot of suction dredgers are spending the summer camping out in the valley."

"I'm looking for the lode source of the gold," I said, grandiosely sweeping my arm around at the mountains. They wished me luck, but obviously, having a nugget in the pan was more satisfying to them than tromping around barren hillsides with the wildflowers. The next day and many days after that I hiked into the area, investigated the abandoned cabin for a mining dump, and picked up veined rocks to assay. I got to know the geology, and wasn't swayed by finding veins of antimony and lead. The gold placers originated in these hills and their source was the goal of my search.

As the metamorphic rocks gave up their secrets, I began to recognize graphitic and mica schist beds rich in sulfides. Along their laminations pyrites and arsenopyrites metallically glinted. I loved to hold up a freshly broken piece and smell the sulfides. Perhaps these were the source of the gold. Their weak gold content could have been remobilized by heat from a nearby magma source. Granite was everywhere, but it didn't create any contact effects where the molten rock had intruded the schist—not a good sign. I reread the reports by the professionals. They seemed to think that either the intrusives were involved or a stratabound occurrence within the bedding of the schists was the gold reservoir. They published a two-inch thick report and walked away from the area without finding the lode source.

THINGS TO DO WHEN PROSPECTING

Each time I prospect an area that has potential, I hike the contour of the hillsides well above the creeks and collect soil samples to have analyzed as part of a geochemical survey. If a deposit exists higher up, it will leak arsenic and gold into the soil downslope in a fan shape. Using a shovel, I dig down below the humus layer (the "A" horizon) into where the soil is free of roots (the "B" horizon) and put some soil into a sample bag. This zone is where the downslope migration of elements is concentrated, and an assay of it will show what minerals are in the area. Digging deeper to where the rocks and bedrock begin (the "C" horizon) only tells me about the bedrock composition in that immediate area so it's not as useful.

When I come across good looking rocks, I bag them, as well. "Pick out the best rocks for assaying," a geologist once told me. "That way you begin with the best the area has to offer."

Most discoveries today are based on soil sampling and heavy mineral panning of stream drainages. Since I already knew the creek had placer gold, I concentrated on searching the hillsides for the source. If an assay indicates an anomaly (a higher than usual amount of a mineral) in the soil, I can return to the site and do more extensive testing. Getting an assay report with a string of gold values in the parts per billion range in the soil is a modern form of discovery.

After investigating the higher mountains, I began camping on the placer tailings in the valleys. There I could see the exposed bedrock along the banks and trace the trend of the bedding. I found where the granite peeked out of the hillside and hadn't been mapped. I sent in more soils and rocks, and got a few hints to follow up. Now I'm near the end of my investigations. Every mountain has been climbed and sampled. The bedrock outcrops don't hold any more secrets. There's one last hillside, located near the headwaters of the creek, where the bushes promise shelter for bears sleeping after a fill of blueberries. I need samples from a gray and yellow altered zone between the granite and the sulfide beds in the schist. Will it be in this last, distant and slightly scary place that I finally solve the question of the source of the gold?

SUMMARY

Discovery is a methodical process. A prospector starts with some hint of mineralization: an old report, a known mining district, or placer gold occurrences. The time comes to camp out and get to know the area. I plan on visiting a prospect a minimum of five to ten times–usually over a year or two. My original ideas are checked out and refined or rejected. My eyes become sharper. I center in on particular places upslope or upvalley, beyond where the placer occurred. Gold can be eroded from a distant source, captured by a stream, the land tilted and uplifted, and the gold captured by another drainage. Soil sampling and rock analysis are done to confirm or abandon a likely idea.

After spending time on a promising area, really getting to know the rocks, and following up all the possibilities, my chances of a discovery greatly increase. My legs are stronger, I feel comfortable in that neck of the woods, and I've completed the most thorough prospecting possible. At times I have to admit that I've gone "deep enough" and walk away. However, I'm in this pursuit against all the odds, for that electrifying moment of discovery when the rocks reveal their treasure.

4. STAKING YOUR CLAIM

Suppose you are out taking a day hike in a wooded valley when you part some brush and find chunks of white feldspar and quartz crystals, indicating that a gem-bearing pegmatite lies nearby. Or perhaps you see nuggets in your pan while panning a remote creek. Could you check the land status, stake out a mining claim, and record your find so you could retire on its income or sale? Claim staking is a simple process that rockhounds and recreationalists should understand, because knowing how to stake a claim in the field, research the legal status of the land, and record a claim so further prospecting work can be done legally are essential to the success of every prospector.

Fred Rynerson described finding a pegmatite in Southern California in *Exploring and Mining for Gems and Gold in the West* (Naturegraph Company, 1967). Not having the materials with him to make out a claim, he went to get paper and pencil. But when he returned, the claim had been staked by someone else. Carrying the basic supplies for claim staking may mean the difference between owning a mineral deposit and losing it.

POSTING AND FLAGGING YOUR CLAIM

A valid claim has to have a claim notice posted on its northeast corner with a sketch of the claim, a description of its location, the date of the discovery, and the claimant's name and address. A plastic bag, jar, or film canister is usually used to protect the claim notice from weather. In addition, the boundaries of the claim must be marked with flagging, and monuments or posts placed at all four corners. On federal land, claims also require a discovery post or cairn where the actual discovery is located. A detailed description then has to be filed with the BLM (for claims on federal land) or state lands office (if it's state land), and the original claim recorded at the district recorder's office.

A claim post in an alpine area is easily visible. This is a lode claim above a gold placer creek. Rock cairns may also be used in a treeless area.

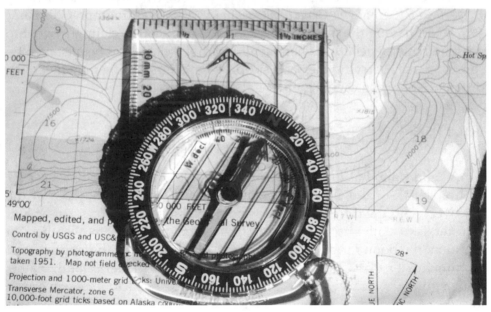

Topographic maps have the magnetic declination for the area shown at the bottom of the map. This orienteering compass is adjusted so that when the needle points to magnetic north, the arrow points to true north–corresponding to the north-south orientation of the map.

A claim staking outfit includes compass (and knowledge of what the magnetic declination is for the prospect area), a topographic map of the area, claim forms, a pencil, flagging, and an ax (not always required) to cut corner posts. I used to pace off distances, but now I use a "hip chain," a counter attached to my belt that spools out a thin thread to measure the distance walked.

To make a straight line I set my compass with a bearing—for State of Alaska claims either north-south or east-west. An orienteering compass has an arrow that can be set to correct for the magnetic declination. In my area the declination is 30° east, so I turn the direction arrow 30° to the west while the needle points to magnetic north. The direction arrow now points true north when the magnetic needle is lined up with magnetic north. Each topographic map has a diagram showing the declination for its particular area.

In most states the land open for mining claims is federal land under the control of the Bureau of Land Management (BLM). On federal lands claims are either lode (600 feet by 1,500 feet) or placer (660 feet by 1,320 feet). Placer claims can be up to a maximum of a 160 acres based upon the number of people included in an association of up to eight people, times twenty acres for each person. A one-person federal placer claim would be 660 feet by 1,320 feet or 20 acres. Federal claims are situated by the discoverer to cover the vein or creek, whatever its orientation. In Alaska about a third of the land is owned by the state, which allows claims (no difference between lode or placer) to be 40 acres, or 1,320 feet by 1,320 feet, in north-south and east-west directions.

CLAIM FEES

Not only is Alaska more generous with the amount of land in each claim, it is also less expensive. Each federal claim requires an annual payment of $100 plus recording fees; each Alaska claim has an annual $25 fee plus recording fees. Recording office fees are separate from the mining fees and are a one-time cost. The annual Alaska payment remains in effect for the first five years of the claim, and then goes up to $50 a year for five years, and eventually tops out at $100 each year. The incremental schedule for Alaskan claims allows a prospector or miner time to assess the deposit, begin mining it, or interest a company in developing it. In addition, the State of Alaska sends each claim owner a list of their current claims shortly after August 31, the end of the mining year, with the option to cross out any claims the owner wants to abandon before paying the annual rental fee for the next mining year. The federal government doesn't send out such a notice.

While doing volunteer work in the State of Alaska Division of Mining office, I helped file hundreds of claims. New discoveries and renewed interest by mining companies and individuals had led to massive numbers of stakings. This caused a pile up of paperwork, which a few of us offered to file away. Erik Hansen, a former expert with the division, whom mining companies hire to research land ownership in areas they are interested in staking, was there to coordinate our work and give advice on knotty problems, and he promised to train us in the finer points of land status research as part of our volunteer time. Katy, a former BLM employee and fire dispatcher, was also helping file claims.

"What about these claims?" Katy showed a batch of papers to Erik. "They were staked on August 30th. Shouldn't they have waited until September first?"

"They wasted a lot of money paying for the last two days of the mining year," Erik told us. "They probably think they've already paid for the next mining year. These claims may be open again."

"I had a similar experience last summer," I told them. "I was worried someone would stake a vein I found, so I put in the claims in July, even though I knew I would have to pay the same fees again after September first. It was worth the peace of mind."

TOPOGRAPHIC MAPS

The most important section on a claim form is the legal description of the claim that ties it down to a particular spot. Topographic maps at the 1:63,360 scale have township and range numbers shown along the sides, and within each township the thirty-six sections are numbered.

In all states the legal description of the land begins with the meridian, the zero point from which land in that area is measured. Alaska has five meridians, so a land description that omits the meridian may be in any

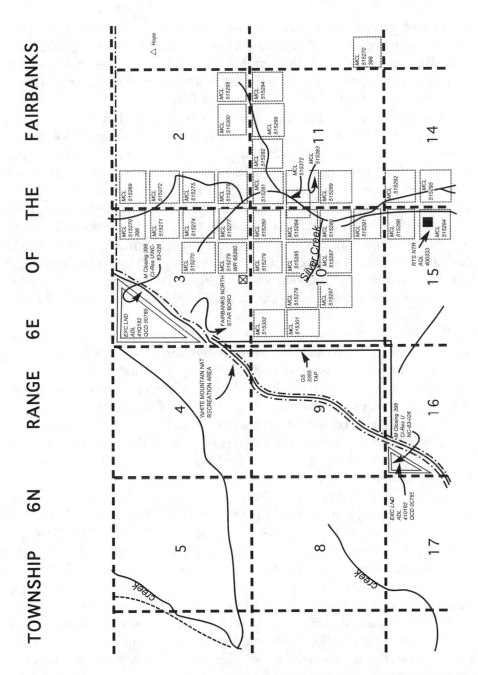

A sample land status plat. Land status plats are available in Alaska for checking open lands, ownership, and mining claim locations. Each status plat is one township (a six-mile square). Supplemental plats for sections (one mile by one mile) are done for areas with complicated ownership.

one of five places. The Fairbanks Meridian runs north-south through an "initial" marker point on a hilltop near Fairbanks. From that point the townships are numbered north and south, and ranges east and west. On the edges of a topographic map these are shown as, for example, T2N for Township Two North and R4W for Range Four West.

A township is a square six miles on a side, and it is divided into thirty-six numbered sections, each of which is one mile square. T2N starts six miles north of the Fairbanks "initial point," and extends to twelve miles farther north. Near cities the township lines have been cleared and can be seen stretching north-south and east-west. Brass monuments are placed at the corner of each section and quarter section (halfway between the section markers).

It helps to remember that a mile is 5,280 feet, a half-mile is 2,640 feet, and quarter-mile is 1,320 feet. Since an Alaskan claim is 1,320 feet on each side, and the federal placer claim is 660 feet by 1,320 feet, a State of Alaska claim is one-quarter of one quarter of a section, or forty acres, and a federal placer claim is one-half of one-fourth of one quarter of a section, or twenty acres.

Each one-mile square section is divided into sixteen quarters–four large ones and four quarters inside each larger one. Each quarter is described using north-south and east-west directions. Begin with the claim itself. Is it in the northwest, northeast, southwest, or southeast quarter of the quarter section? One of the easiest claims to describe is the one in the upper right-hand corner. It's in the northeast quarter of its quarter section. In the section, it's in the northeast quarter. A state claim in the area about ten miles northwest of Fairbanks might be written as "the NE 1/4 of the NE 1/4, Section 10, T2N, R4W, Fairbanks Meridian."

Since federal claims are one-half the size of an Alaskan claim, a legal description would add "North 1/2" (or "South 1/2" or "East 1/2" or "West 1/2") of the NE 1/4 of the NE 1/4, Section 10, T2N, R4E, Fairbanks Meridian. This description would give you a federal placer claim 660 feet by 1,320 feet. Most creeks don't run exactly east-west or north-south of course, so you may just list all the quarter sections the claim is found in, and a map on your claim form would show the true location of the claim.

My first claim descriptions were inaccurate, and nowadays I often come across claims in the woods that lack exact descriptions, making them legally invalid. It's important to practice describing a claim location on a topographic map until it can be done correctly. Plus, I always double-check all my legal descriptions before submitting my claims.

Revised 7/00
DNR 10-162V
Traditional

STATE OF ALASKA
Department of Natural Resources
STATE MINING NON-MTRSC LOCATION NOTICE / CERTIFICATE
(TRADITIONAL CLAIM ONLY)
USE THIS FORM ONLY FOR CLAIMS BASED ON "GROUND STAKING" ON STATE LAND.

CLAIM NAME/NUMBER: **Old Dog #20**   CREEK NAME: _____
                                                                (optional)

DATE OF LOCATOR'S DISCOVERY: **10/24/92**   DATE ORIGINAL LOCATION NOTICE POSTED: **1/16/93**
                              (month/day/year)   (month/day/year)

THIS CLAIM IS (maximum area = 1320' x 1320'): **1320'** ft. long in a N/S direction and **1320'** ft. wide in an E/W direction.

THIS CLAIM IS LOCATED IN **Fairbanks**   RECORDING DISTRICT. (complete as many lines below as apply to this claim):

Fairbanks	Meridian: Township **2N**	, Range **1W**	, Section **8**	, Qtr/Qtr Section **SW1/4**	of **NW1/4**
_____	Meridian: Township ____	, Range ____	, Section ____	, Qtr/Qtr Section ____	of ____
_____	Meridian: Township ____	, Range ____	, Section ____	, Qtr/Qtr Section ____	of ____
_____	Meridian: Township ____	, Range ____	, Section ____	, Qtr/Qtr Section ____	of ____

LOCATOR (The locator is the owner. Print the Name & Address below where correspondence should be sent):

1. Owner's Name: **Roger McPherson**
Mailing Address: **111 Juneau Drive**
City, State Zip: **Fairbanks, AK 99717**
Contact Phone: **(555) 123-4567**

ADDITIONAL LOCATORS/OWNERS (Please print):

2. Owner's Name: _____
Mailing Address: _____
City, State Zip _____
Contact Phone: _____

3. Owner's Name: _____
Mailing Address: _____
City, State Zip _____
Contact Phone: _____
(Use extra page if necessary)

ALL OWNERS OR THEIR AGENTS MUST SIGN:
I hereby certify that, on the date above, a location notice was posted on the monument at the NE corner of this claim, and to the best of my knowledge, in accordance with applicable State statutes and regulations.

1. Owner /Agent Signature: *Roger McPherson*
                         (X) OWNER   ( ) AGENT
Agent's Name (print): _____

2. Owner /Agent Signature: _____
                         ( ) OWNER   ( ) AGENT
Agent's Name (print): _____

3. Owner/Agent Signature: _____
                         ( ) OWNER   ( ) AGENT
Agent's Name (print): _____
Attach an extra sheet for Additional Owners and Owner / Agent Signatures if necessary.

CLAIM SKETCH
(see instructions on reverse):

	NW/NE	NE/NE
NW	SW/NE	SE/NE
SW	SE	

Scale: 2" = 1 Mile

= Size of traditional STATE mining claim (1320' x 1320', or 40 acres. [E.g. SE QTR of NE Qtr]

= One section at 1" = 1 mile. (1"63,360) scale.

Optional *In place of the claim sketch to the left, a separate map showing the location of this site is:*
☐ Attached to this Certificate.
☐ Attached to the Certificate for the following claim. _____

Recorder's Office Use Only:

Div of Mining, Land & Water Use Only:
ADL_____.

RECORD THE ORIGINAL (WHITE) PART OF THIS FORM AT THE DISTRICT RECORDER'S OFFICE. MAIL OR DELIVER THE YELLOW COPY (or photocopy) WITH THE FIRST ANNUAL RENTAL PAYMENT AS INSTRUCTED ON THE BACKSIDE OF THIS FORM.

A filled-out sample State of Alaska Mining Claim form. An abandoned State of Alaska Mining Claim (40 acres or 1,320 feet by 1,320 feet) has the basic information required for staking and recording a valid claim. Correctly describing the location of the claim takes practice.

EXAMPLES OF METHODS OF MONUMENTING MINING CLAIMS IN CALIFORNIA

Drawing of an ideal lode mining claim
(Metes and Bound survey method) in California
(Cal. Pub. Res. Code, Chapter 4, Sec. 2316).

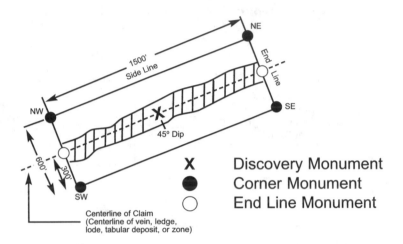

X   Discovery Monument

⬤   Corner Monument

◯   End Line Monument

Most State laws require conspicuous and substantial monuments for all types of claims and sites.
NOTE: Other states have other requirements for monuments. Other monuments can be used in California as long as they are conspicuous and substantial. However, it is BLM policy to not use perforated or uncapped pipe as a monument.

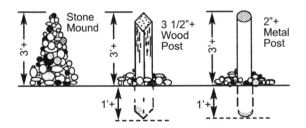

A page from the Bureau of Land Management booklet, Mining Claims and Sites on Federal Lands, *illustrates federal claim types, sizes, and descriptions. Federal lode and placer claims are slightly different in size.* Courtesy of Bureau of Land Management

METHODS OF DESCRIBING PLACER MINING CLAIMS AND MILL SITES

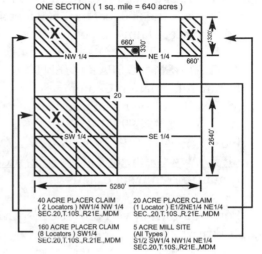

MOUNT DIABLE MERIDIAN (MDM)
T.10 S., R. 21 E., Section 20

ONE SECTION (1 sq. mile = 640 acres)

○ Location Monument

X Discovery Monument

40 ACRE PLACER CLAIM
(2 Locators) NW1/4 NW 1/4
SEC.20,T.10S.,R21E.,MDM

20 ACRE PLACER CLAIM
(1 Locator) E1/2NE1/4 NE1/4
SEC.,20,T.10S.,R.21E.,MDM

160 ACRE PLACER CLAIM
(8 Locators) SW1/4
SEC.20,T.10S.,R.21E.,MDM

5 ACRE MILL SITE
(All Types)
S1/2 SW1/4 NW1/4 NE1/4
SEC.20,T.10S.,R21E.,MDM

Drawing of a section of land showing types of placer mining claims (PMC) and a mill site (MS). The legal descriptions method is based on the U.S. Public Land Survey.
Courtesy of Bureau of Land Management

A valid mining claim has a description of the claim posted at the northeast corner, posts or cairns on all four corners, and clearly marked boundaries. The author puts in ten to twenty claims a year.

CLAIM FORMS

Claim forms are available with spaces for all of these descriptors. Copies of Alaska's claim form (useful for federal claims if you scale your claim down to the 600 feet by 1,500 feet size for lode and 660 feet by 1,320 feet for a placer claim) can be obtained from:

THE ALASKA DEPARTMENT OF NATURAL RESOURCES
Public Information Center
550 West 7th Avenue, Suite 1260
Anchorage, AK 99501-3557
(907) 269-8400 / (907) 269-8901 Fax
http://www.dnr.state.ak.us/pic/index/htm
E-mail: pic@dnr.state.ak.us

For an informative booklet called *Mining Claims and Sites on Federal Lands* write to:

BUREAU OF LAND MANAGEMENT
1849 C Street NW, LS-510
Washington, DC 20240-0001
Phone: (202) 452-0350

Examples of federal claim forms and detailed procedures for staking claims in Nevada can be found in *Mining Claim Procedures for Nevada Prospectors and Miners*, a booklet published by the Nevada Bureau of Mines and Geology (*Special Publication 6* by Keith G. Papke and David A. Davis). Write to:

THE NEVADA BUREAU OF MINES AND GEOLOGY
Mail Stop 178
University of Nevada
Reno, Nevada 89557-0088
Order online at http://www.nbmg.unr.edu, click on "Publication Sales"
    or call (775) 784-6691 ext. 2

CERTIFICATE OF LOCATION

LODE MINING CLAIM

TO ALL WHOM IT MAY CONCERN:

The Undersigned hereby certifies that he has caused to be located the _____
_____ Lode Mining Claim in the following quarter section(s):

1/4	Section	Township	Range	Meridian
___	_____	_____	_____	_____
___	_____	_____	_____	_____
___	_____	_____	_____	_____

in _____County, Nevada, on the _____day of _____, 20 ____.

The name and mailing address of the locator is:

_____

_____

_____

The Claim is approximately _____feet long and _____feet wide, such that _____feet is claimed in a _____direction and _____feet in a _____direction from the point of discovery (monument of location), at which the Notice of Location was posted, together with _____feet on each side of the monument of location and center line of the Claim. The general course of the lode or vein is from the _____to the _____direction.

The number, location and markings on each side center or corner monument are as follows:

	Location	Markings	Description
No. 1:	_____	_____	_____
No. 2:	_____	_____	_____
No. 3:	_____	_____	_____
No. 4:	_____	_____	_____
No. 5:	_____	_____	_____
No. 6:	_____	_____	_____

As erected on the ground, each side center or corner monument is marked as described above by _____ (e.g. metal tags, paint on posts).

The work of location consisted of making a claim map as provided in NRS 517.040.

Dated this _____day of _____, 20 _____.

SIGNATURE OF LOCATOR:

BY: _____
         (Agent)

* * * * * * * * * * * * * * *
*   RECORDER'S STAMP
*
*
*
*
*
*
*

Suggested Form – Nevada Department of Minerals
Nevada-Lode Certificate of Location - NRS 517.050

A blank federal lode claim form from Nevada.
Federal claims have a $100 annual payment.

Many mining claims don't have an accurate sketch and description posted at the northeast corner, clearly marked boundaries, or a valid mineral discovery. Some people simply look at a map and feel that they don't need to go out to the place before staking a "paper" claim. They make out an accurate claim form, send or take it to the BLM or state office, pay their fees, and record it at the recorder's office. There's nothing on the ground, no notice in the northeast corner, and no flagging or signs to indicate the boundaries of the claim. Unless another prospector checked the land status plats, they wouldn't know the area was claimed. Canada has mining claim inspectors who check claim boundary markings and corner posts, but in the United States, claims aren't inspected, making paper claims difficult to challenge in court.

FOLLOWING UP ON CLAIMS

The mining laws require that claim owners go out and check their claim lines on a yearly basis. Many don't do this and they forget exactly where their old markers are. The other extreme is the person who blazes each tree with an ax, leaving scars that will always be there. I use pink surveyor's flagging to mark my lines, and put up the required four-inch by four-inch post or cairn at each corner. If I abandon the claim, I walk back along the lines and take out the ribbon and posts, leaving nothing behind. I also put up a sign wherever people are likely to enter the claim area. Berry pickers, hunters, fishers and hikers can still use the mining claim—I only have the right to the minerals.

"My very own mining claim!" I hear people say, a gleam in their eye. "I'll build a summer cabin and have the perfect retreat." This practice has been stopped. A mining plan must be submitted for approval before any development can take place, and any disturbance must have a reclamation plan. Every winter I do an annual mining plan (an APMA, or Annual Placer/Lode Mining Application) for the State of Alaska and pay a $100 fee. This plan is sent to many agencies, and they check on whether I meet wetlands guidelines, satisfy fuel storage laws, or have any access problems. Field personnel from the State of Alaska Division of Mining schedule visits to the claim and discuss any problems they see in my plan. The Army Corps of Engineers are also involved in the permitting process whenever my claims are on wetlands.

Before going into an area to look for minerals, a search of land status will show who owns the land and what restrictions apply. The BLM has public rooms where information on land status and federal mining claims can be found. BLM's computer system uses townships and ranges to research a particular area. The majority of mining discoveries in the United States were made when the federal government owned the land. Some of these claims were patented while others received a case file number, but were not patented. If federal mining land is patented, it becomes private land, and unlike a regular mining claim, which requires annual assessment work to maintain its validity, patented land is subject to property taxes, and may be developed into any real estate use. So the BLM records have to be checked to find out if a federal claim is still active or is open for staking.

On a broad scale the United States Geological Survey, USGS, has land status ("surface management status") maps done by the Bureau of Land Management, BLM that outline the ownership for land in some western states. These maps can be obtained from USGS map centers, and they show that many lands are closed to mineral entry. In Alaska, for instance, I can't stake claims in national parks, wildlife refuges, wild and scenic river corridors, or state recreation areas. So I look for open federal land and state land that isn't closed to mineral entry.

From the time I discover and stake a State of Alaska Claim, I have forty-five days to record it. This is a time when an unethical person might find my posted claim notice in the field, or see a copy of the claim on file. They could overstake my claim, back-date their claim posting date, and say I was the claim jumper. However, the official date the recorder puts on a claim can't be beaten in court, so getting to the recorder's office soon after making a discovery is a priority. Another view holds that waiting until just before the end of the forty-five-day period (ninety days for federal claims) before filing and recording prevents someone from seeing your claim become public and closes off the option of them back-dating their claim. I have been overstaked, so beware of these loopholes. Unfortunately, such disputes usually must be settled in court.

To keep a claim active the claimant must pay the annual fee and, in addition, do $100 worth of annual labor for each claim (this is the same for both federal and State of Alaska claims). An Affidavit of Annual Labor must be filed, giving a legal description of the claim, the work performed (assays, trenching, testing, etc.), and the total amount of money spent.

DOM 20-83
(revised 08/00)
DNR 10-84

Read instructions on back
before completing this form.

AFFIDAVIT OF ANNUAL LABOR
FOR MINING CLAIMS, LEASEHOLD LOCATIONS,
OR MINING LEASES
(Type or print)

United States of America)
) ss:
State of Alaska)

1. This affidavit of annual labor is for the assessment year which ended at noon on September 1, 20 **02** .

2. **NAMES OF MINERAL LOCATIONS:**   **ADL NUMBERS:**

| Old Dog 1-4 | 526184-187 |
| Old Dog 12-23 | 540564-571, 543712-715 |

(Attach additional sheets if necessary)

3. These claims are located in the following Meridian(s), Township(s), Range(s) and Section(s):
   Fairbanks Meridian, Township 2N, Range 1W, Section 8

4. These claims are recorded in the **Fairbanks** Recording District.

5. Current owner's name & address (Correspondence sent to):
   NAME: Roger McPherson
   Address: 111 Juneau Drive
   Fairbanks, AK 99717

6. Excess work in the amount of $ _____ is to be applied from previous years.

7. Work was performed on the following dates: May 15,16,17,18,19,20,21,25;June 24,25,29,
   July 1,4,6,7,28,30; August 2,16,17,18-1996

8. Number of person-days worked: 33

9. Description of work performed and improvements made are as follows: Extension of soil grid, soil sampling
   using hand held auger and track mounted auger, compilation of data, analytical costs.

10. Value of work performed not including claim maintenance: $ 17,259.00

11. Name(s) & address(s) of person(s) who did the work:
    NAME: Joe Smith (Consulting Geologist   NAME: Jane Smith, Laborer
    Address: 1212 Gold Mine Lane   Address: 3434 Diamond Lane
    Anchorage AK 99999   Anchorage, AK 99999

12. If the labor was performed by someone other than the owner or the owner's lessee, the amount paid for the work and
    improvements made was $ _____ and _____ made the payment.
    (Name)

13. Amount of any cash payment made to the state instead of performing annual labor: $ _____ .
    Date of payment: _____ . (NOTE: Payment must be made before September 1. **This
    affidavit of annual labor must still be timely recorded.**)

14. I, Roger McPherson _____ , swear under penalty of perjury that the foregoing is true.

    x _Roger McPherson_
    (Signature of Affiant)

    | | (Recorder's Use Only) |

Subscribed & sworn before me
this _____ day of _____ , 20 ___
Signature of notary _____
My commission expires _____

*A filled out sample Affidavit of Annual Labor form. Federal mining claims and
State of Alaska claims require an Affidavit of Annual Labor. This lists the work
done to satisfy a $100 a year expenditure for each claim to maintain the claims.*

On State of Alaska claims, a large outlay in one year can be spread out to cover future years, but excess annual assessment work on federal claims cannot be carried over. Blank Annual Labor forms can be obtained from BLM or the State of Alaska, and these provide spaces for the required information. People holding no more than ten federal mining claims, however, can file a small miner's exemption, or "Maintenance fee payment waiver certification," and avoid the annual labor costs. Filing deadlines are different for all of these payments and forms, and missing a deadline will constitute abandonment of the claim.

SUMMARY

As I helped file claims at the Division of Mining, I noticed that some filers were "mountain claimers," who claimed large blocks of country, and others who only filed one or two claims. Most mineral deposits are not extensive enough to warrant claiming a whole mountain, but as long as you have a mineral discovery on every claim, you are allowed as many as you want. An untouched, gold-bearing creek might need ten or twenty claims to cover the placer deposit. This runs into a lot of money for fees, and Herculean efforts to mark every claim properly. Then, each year annual labor affidavits have to be filed.

In the old days, miners were only allowed their discovery claim and one other on the creek. Now, a mining company generally wants to claim a large block to protect its interests. When the Fort Knox Gold Mine near Fairbanks was discovered, for instance, small claim holders in the surrounding area did well selling their claims as the mine consolidated its claim block.

Prospectors, rockhounds, and others who spend time outdoors are smart to be prepared in case of a mineral discovery. Knowing the status of land in the area; having some compass skills and map-reading ability; and carrying claim forms or paper and pencil, flagging and a plastic bag or film canister to protect the claim notice enable a person to make an on-the-spot claim. Some discovery stories end sadly because the discoverer didn't know how to properly establish the claim and maintain it over a period of time. But with a little preparation and practice, anyone can stake a claim and maintain legal ownership of a mineral deposit.

5. FORT KNOX: THE ONE THAT GOT AWAY

One of the first "lost prospect" stories I heard was on a bus with a group of exploration geologists while touring the Paradise Peak Gold Mine near Gabbs, Nevada. Just after we turned off the main road into the mine area a man wearing a cowboy hat and a large belt buckle stood up in the swaying bus.

"I drove by this place for years," he began. "Paradise Peak stood out as a red, yellow, and orange mountain just off the road. We all sampled around it." He sadly shook his head. "There was even an antimony mine on one side of it. Then, one day some guys went to the top of it and got some more samples."

"And you've been kicking yourself around the sagebrush ever since," his seatmate interjected.

"A $400 million mine and I drove by it for ten years." The Nevada geologist looked at his silent audience. "It was a mine from the first day of operation. The whole mountain was mineralized...just not the edges where we sampled." He sat down then, and I remember thinking what a sad story that was.

Probably every geologist on that bus had a similar experience haunting them, but no one was going to admit it. The world class Fort Knox Gold Mine near Fairbanks, Alaska, is my personal "one that got away" story. When I'm out on a remote creek and stop to enjoy the view, a nagging reminder of this fiasco intrudes and I wonder if I serve some greater purpose. There is a learning curve in all things, but losing a mine is a hard lesson.

There are signposts that advertise the presence of a mineral deposit. Mountains colored red from oxidized pyrites have long been sampled and drilled to find the copper or lead-zinc deposit below. An antimony or mercury mine suggests an ancient hot springs system above a gold or silver deposit. Similarly, veins of lead, zinc or silver may indicate greater bonanzas at depth, and gold placers in a creek suggest lode deposits in the nearby mountains. Such was the case with the Mother Lode of California, once the placer miners got beyond the initial rush for creek nuggets, and with the great Homestake Mine of Lead, South Dakota, where prospectors followed the Deadwood gold placers to their upstream source at Lead.

PROSPECTING FOR LODE SOURCES NEAR FAIRBANKS, ALASKA

Fairbanks began in 1902 when placer-rich creeks were discovered scattered around three spruce- and birch-covered domes. Pedro Creek, named after Felix Pedro the discoverer, and Goldstream, Gilmore, Cleary, Wolf, Fairbanks, Fish, Cripple, Ester, Engineer and other creeks made men rich from their gold. A few small hardrock mines on veins and sheer zones were worked but the schist of the area was a poor host for gold lodes. The thinking was that erosion had removed miles of rock from the region and eroded the productive granite domes to their barren cores. "Rich placer, poor lode," say the old prospectors. The reverse situation is also possible: a small placer may indicate the existence of a large, undiscovered lode source.

When I began studying prospecting, the books by economic geologists emphasized the importance of magma in the formation of mineral deposits. Melted crystalline mantle and crustal rocks in the subduction zones, where oceanic plates slid under the edges of continents, formed these igneous rocks that could host copper, tungsten, molybdenum, tin, or gold deposits.

J.B. "Beaver" Mertie, one of Alaska's tireless USGS geologists, suggested that areas around granite should be the target of prospecting in Interior Alaska. Following up on his advice, I decided to examine all the known granites in the Fairbanks area as a class project.

One sunny day I visited Dave, a placer miner, working off one side of Gilmore Dome in a steep creek that was cluttered with granite boulders. He and his partner John were just pulling up the matting in their fifteen-feet single-channel sluice box. A fine glitter of gold covered the green Astro turf.

"Look at this one!" Dave said, holding up a bright nugget. "Gold is everywhere in this decomposed granite," he exclaimed. He grabbed a gold pan and scooped up some of the granular material from the creek bed. "We love sluicing this stuff. It runs through cleanly without a lot of silt." He tilted the pan to show me the speckled bottom. "I've never had such a good time mining." Like two kids, they continued with their clean-up.

I visited other granite exposures encircling Fairbanks and claimed one that had pyrites and arsenopyrites along its edge. With the different rocks I'd collected, I made a display of the varieties of granites. Pedro Dome's was diorite to granodiorite, a fine-grained, salt-and-pepper texture. Richer in quartz, Gilmore Dome, a quartz monzonite, had creamy feldspar crystals and a ground mass freckled with biotite flakes.

The tops and smaller satellite bodies associated with granites have more potential for mineral deposits than the big plutons that make up mountains like Pedro Dome or the core of the Sierra Nevada Mountains in California. Gilmore Dome had a satellite body of granite shaped like a half-moon, and I frequently drove the back roads near this to look it over. The creek below it had had a small gold placer, and geologist Mertie had written about a prospect on this hillside where gold and bismuth were panned from the crumbly granite. I hiked along a powerline road, hammering at samples of the speckled rocks.

A mossy hillside with stunted black spruce covered the rest of the half-moon granite. I knew a few claims covered part of the area. One nearby claim owner hadn't been very friendly when he found my car parked on the road, and he had a cable blocking the access. I did want to find that crumbly granite prospect, but walking on deep moss is tiring work. Hadn't I seen enough? I left with my samples and focused on another area above Treasure Creek, where a rich placer stretched downstream for three miles.

A few years later, a CAT (bulldozer) driver with a geology background happened to scrape the hillside near the center of the half-moon shaped granite. He saw thin quartz veins crisscrossing the exposed rocks. Cross-cutting veins in any rock are cause for attention. These eventually became Fort Knox, one of the largest gold deposits in granite ever found.

THE FORT KNOX GOLD MINE

I refrained from telling my story during a field trip to Fort Knox a few years later. For the culmination of an ore deposits class our professor had scheduled the field trip, and as we followed the well-graded dirt road that dipped across the low hills near Gilmore Dome before curving down to the mine buildings, we could see open pit mine benches cutting across the hillside. After the bus carried me and the group of young geologists through the security gate, we were given yellow hard hats and safety glasses to wear.

The chief geologist directed the bus past bays where 150-ton haul trucks were serviced, and past huge buildings containing the crushers and milling apparatus for leaching the rock even during the dark winter when temperatures are -50° F. We headed for the central pit and an inactive area beside the exposed pit wall.

"We are mapping each level as the cut goes down," the geologist explained. He spread out his maps and we gathered around. Colored lines on the map crossed like bird tracks on a beach. In the bright sunshine, the map's notations faded into incomprehensible marks.

We were finally released to search for specimens. The geologist pointed out major veins, faults and shears. Rock hammers bounced off the resilient granite. Below us an excavator filled a million-dollar haul truck, one of nine that work all day and night year round.

I looked up at the saddle where I had once stood. Drill pads and roads covered the hillside and merged into the pit walls. I had picked blueberries up there among the barren granite. I had walked away from a great discovery. Sadly I joined the other students and filled a white sample bag with rocks.

A few months later I returned for a more complete tour of the mine with a group from the Chugach Gem and Mineral Society of Anchorage and some of my University of Alaska Mining Extension buddies. Our long string of cars stopped at the mine overlook for pictures and a brief talk. I found myself explaining the role of granite in forming gold in the Fairbanks District. I swept my arm over the half-moon of the Fort Knox granite now exposed as a white gash across the hillside.

"You've got to examine all areas of a granite," I advised the group. "One end may be barren, but some other area may contain a concentration of minerals." I thought: *I had stood right on the edge of the mine and I didn't walk the extra distance to explore it.* "This deposit was hinted at in an early

USGS report. Prospectors panned gold from the disintegrating granite." *Next time I'll act on what I read in old reports.* "They found criss-crossing quartz veins in the granite." *If you don't walk on it and see it, you won't discover it.* "The area only had a few claims." *Don't be put off by existing claims or a protective claim owner.* "They have 50,000 acres under claim now."

At the modern white and blue mine building we were ushered into a meeting room for an introduction. "I'm Alissa, your tour guide," a young employee told us. "The Fort Knox Mine has a twelve- to fifteen-year life expectancy. We began construction in 1995 on the $375,000,000 plant, which contains crushers, ball mills, leaching tanks and detoxification circuits to neutralize the leached rock before it's discharged into the tailing pond."

She showed slides of the past and projected production of 300,000 to 350,000 ounces of gold a year. Each day the plant produces 1,000 ounces of gold from rocks containing only 0.025 to 0.035 ounces of gold per ton! The mine contains 4.1 million ounces of mineable reserves, and they are looking to expand that to 5 million ounces.

Toronto-based Kinross Gold Corporation merged with Amax Gold, a subsidiary of Cyprus Amax Minerals Company, and acquired Fort Knox. Kinross produces more than one million ounces of gold a year from its

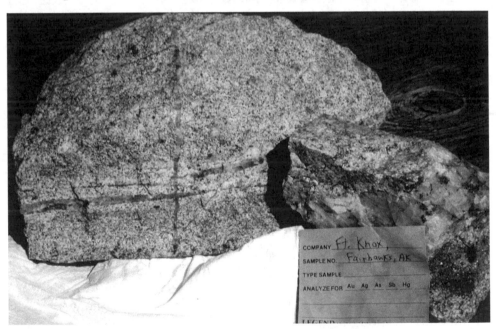

Cross-cutting quartz veins in granite from the Fort Knox Gold Deposit near Fairbanks, Alaska are unmistakable signs of mineralization. White-colored alteration along the edges of the veins is from the introduction of albite feldspar.

varied mines, and expects to boost Fort Knox production and reserves. Last year Fort Knox spent $170 to produce each ounce of gold, but if the mine debts are included, the real costs rise to $342 per ounce. With the price of gold at less than $300 an ounce, making a profit is difficult.

After the lecture and slide show we got into two vans and drove past a haul truck that was being repaired. "The trucks are twenty-one feet high," Alissa said. "One driver told me it was like driving a little house from a second story window."

"Do you really work all winter?" one of the group asked.

"Minus twenty is the best time to drive the mine roads," she answered. "We're going up to the West Pit where you can get a good overlook of the operation."

When we got out, I looked around at the rocks. Everyone was watching the pit operation and taking pictures. One pile of recently blasted rocks had beautiful examples of the veins. I grabbed a large specimen to show the group.

"Fort Knox ore contains almost no sulfides and only minor amounts of stibnite and arsenopyrites," Alissa informed us. "It's extremely easy to leach because of the high silica content and lack of sulfides."

Pat, one of my prospecting classmates, had been examining granite around her claims in the Circle Mining District, about 120 miles northeast of Fairbanks. She hefted the rock. "We have a similar granite, which is related to the gold placers in my district," she noted. "This gives me an idea of what to look for."

We headed back to the mine plant for a tour of the mill circuit where 36,000 to 40,000 tons of rock are processed every day. We wouldn't see the primary crusher, which pulverizes one hundred sixty tons in ninety seconds to a six-inch size ("like a coffee grinder," Alissa said), but we did stand in front of the ball mills as they reduced the rock to the fineness of talcum powder. This operation was controlled by a bank of computers and two employees in a soundproof room who watched the swiftly turning ball mills through large windows.

Next was the "tornado in a can," where water was mixed with the powdered rock, and twenty percent of the gold was extracted in a gravity process. Leaching tanks containing a cyanide solution dissolve the rest of the gold from the rock powder ("like sugar in water" according to Alissa) in a short seven-hour period. Burnt coconut shells from Sri Lanka are used in carbon tanks to capture the gold, and then stainless steel screens and electrotwinning (a process similar to electroplating) turn the gold into a sludge that is heated to 2100° F to form a gold bar about the size of a long loaf of bread but weighing eighty to ninety pounds.

After all this we each got to hold a "small" gold bar worth about $94,000 and a Polaroid was taken of us grinning like instant millionaires. Pat got to keep the veined granite specimen to study, and I had a few samples tucked away in my camera bag as souvenirs.

SUMMARY

Now, when I'm out tracing a granite outcrop along a hillside, I think about Fort Knox and what it taught me. No matter how thick the alders or how bad the mosquitoes, I try to be thorough and examine everything. A miner's "keep out" signs don't stop me from hiking into a valley to check out the area. Miners are protective and I respect their operations, but they don't have the right to deny access. I do try to pay more attention to the red flags that indicate potential mineralization, such as the presence of historic gold placers. The element of chance in discovering an "elephant" can be reduced, but finding an economic mineral deposit is never a sure bet–and I've got the story to prove it.

6. Beyond Gold in the Gold Pan: Saving the Semi-Heavies

ANALYZING STREAM GRAVELS WITH A GOLD PAN

With the emphasis on gold, panners often overlook the real utility of the gold pan as a tool in the analysis of stream gravels for a wide spectrum of minerals and gems. When using the gold pan, the heavier specific gravity minerals such as gold, platinum group metals, tungsten, and tin are the most sought after. Yet, with a little more time and effort a good panner can recover eighty to ninety percent of lighter minerals–the semi-heavies–that have specific gravities of 4.0 to 6.0.

While doing a placer evaluation near Nome a few years ago, the huge quantity of garnets clogging the riffles reminded me of why Nome was famous for the ruby sands of its beaches. Abundant garnet in a gold pan will mask other heavy minerals and make it difficult to separate out the gold. However, in most areas garnets are uncommon. Because of their low specific gravity, finding garnets (specific gravity 3.5 to 4.3) or magnetite (specific gravity 5.2) in pan concentrates is an indication of careful panning.

Many minerals with mid-range specific gravities found in placer concentrates give important clues to the types of lode sources in the drainage. For instance, chromite (specific gravity of 4.1 to 4.9), nickel- and cobalt-sulfides (specific gravity of 7.8 and 6.3) along with olivine and pyroxene (specific gravity of 3.2 to 4.4) will direct the search to the "dark" mantle source rocks in the area. A tin-bearing system might have the lighter specific gravity minerals fluorite, tourmaline, and beryllium (specific gravity of 2.7 to 3.3) as well as the heavier cassiterite (specific gravity of 6.8 to 7.1), tantalum (specific gravity of 6.5 to 8.0), and niobium (specific gravity of 5.1 to 6.5) minerals. Chalcopyrite (rare in the oxidizing environment) and copper nuggets (specific gravity of 8.9), sphalerite (zinc with specific gravity of 4.0), galena (lead with specific gravity of 7.6), and arsenopyrite (specific gravity of 6.0) may also be found in concentrates.

EARLY PANNING EXPERIENCES

I first got involved in gold panning when I attended an auction of Harry Francis' lifetime accumulation of placer mining paraphernalia soon after I arrived in Fairbanks, Alaska, in 1970. Looking like Ben Gunn in *Treasure Island*, a wild-eyed Harry stood in the doorway of his broken down trailer, disturbed by all the people and the speed at which his things were being sold off.

"What am I bid for this gold scale and weights?" the auctioneer asked.

In the middle of lively bidding Harry appeared with a small flask of mercury. "Somebody could probably use this too," he offered. "There might be some gold in it."

It seemed like the end of an era to me. Boxes of tools, a collection of broken chairs, and old magazines were quickly sold off. When it was all over and people were leaving, I saw a battered army-surplus 4 x 4 truck off to one side.

"Is that for sale?" I asked. After a brief consultation with the auctioneer, bids were taken. I drove it away for $600 and in it Harry left me his legacy of hip waders, handyman jack, shovel, and metal gold pan behind the seat.

The truck served me well for several winters, even starting at -30°F and -40°F (after I heated it by putting a two-burner Coleman stove under the engine). When I replaced the differential, I found gravel in the case from when Harry had driven up creeks to prospect.

His gold pan was made of heavy metal and was smaller than what I had expected. I tried it out at the creek across from Pedro Monument, a granite monument where Felix Pedro first found gold in 1902, about ten miles northeast of Fairbanks. Gold dredges long ago had worked the area, but the lean gold-bearing gravels in that section of creek are open for recreational panning. A few garnets in the bottom of my pan showed me I was on the right track.

One day I found myself at the headwaters of a small creek, shaded by the trees and cooled by the water. Mosquitoes pricked at my exposed waist, and I was stiff from bending over, but there was the hint of gold in the pan. Harry would have grinned at my slow technique. Another cheechako (slang for a newcomer) had joined the ranks of the gold panners.

Gold panning is a method for obtaining pan concentrates, the heavier specific gravity minerals. Careful panning and examination or assaying of concentrates can give clues to mineral deposits in a drainage system.

PANNING METHODS

My prospecting teacher, Jim Madonna, a professor at the University of Alaska's Mining Extension, presented a paper at the Tenth Alaskan Placer Mining Conference in 1988 in Fairbanks about small-scale mining of flood gold. Flood gold is also known as "flour," "float," and "skim" gold because it does not sink in sand and gravel to form placers. He sampled the top few inches of a gravel bar for flood gold to determine if a small-scale operation could succeed, and his technique illustrates how to use a gold pan to evaluate a creek or gravel bar for gold.

He divided the sandbar into a grid, and used a sixteen-inch pan to process material from about twenty locations, noting the number of colors on surveyor's ribbon that he tied to a rock at each station. He identified a "twelve-color zone" within the grid where more intensive follow-up work could be done with a suction dredge, high-banker, or Gold Screw concentrator. After processing larger quantities from the enriched zone, he concluded that it wasn't profitable to mine this area, even at the small-scale level.

As a boy I remember my father beside a creek in Calaveras County, California, sloshing reddish gravel and muddy water out of a pan. Since then I've learned that a lot of effort is saved by using a "grizzly" or large mesh screen to get rid of rocks before the panning begins. I also use a second screen with one-eighth-inch mesh to further reduce the material. Initially, I'm just looking for colors and analyzing the concentrates in my pan. Nuggets will be captured later in a sluice box or other equipment that can process larger volumes.

Instead of the swirling motion many people associate with gold panning, I use an in-and-out motion. After submerging a pan full of material, breaking up clays, shaking it to classify the heavies toward the bottom, and sweeping out the top layer, I begin a slow series of in-and-out movements. Some books on gold panning reserve this in-and-out motion for the last steps in the panning process.

The pan is kept near the water and tipped so the material in the pan is roughly horizontal with the creek water. Tapping on the sides of the pan and jerking from side to side while the material is kept in suspension stratifies the heavier and lighter materials. The tip of the pan is then dipped into the quiet stream and water enters the pan. As the pan is withdrawn from the stream, the water flows out, carrying the top layer of lighter material off. The edge of the pan is slightly lowered to facilitate the out-flow. Dip in and out to wash off the top layer. This is repeated a few times, and then the pan contents are re-stratified before dipping in and out again.

FINE TUNING GOLD PANNING

"C'mon Big John!" A cheering section surrounded a burly fellow wearing red suspenders in the J. C. Penney's parking lot gold panning contest in downtown Fairbanks. He glanced at his competitors as he rapidly sloshed his muddy pan. Each of us stood by a trough of water furiously shaking, swirling and spilling the contents of our pans.

"I got it!" Big John held a huge nugget aloft after a scant ten seconds.

"Not much of a contest," I grumbled to my neighbor.

Of course, I hadn't started out with a monster nugget in my pan as large as Big John's. The contest allowed each entrant to bring a nugget. Just the dirt was supplied. The weight of his nugget would ensure that it stayed in the bottom of his pan no matter how fast he flung the gravels around. Such contests emphasize that big is better, rather than focusing on the finesse which careful panning requires. In the search for micro-gold and semi-heavy minerals, careful use of the gold pan can open up lode possibilities beyond coarse gold and nuggets.

John Beaver Mertie of the USGS was one of the first to write about the different uses of the gold pan. During his years working in Interior Alaska and in the Southern Appalachians of the United States, Mertie used his gold pan on decomposed bedrock to understand its original composition.

Rather than trying to dig down to bedrock, he would take a sample to the nearest creek and pan it down. Minerals such as the feldspars of igneous and metamorphic rocks are altered to clays; and siltstones and sandstones also break down. However, biotite, quartz, zircon, monazite, and other resistant minerals would be left in enough quantity to reconstruct the original rock type. Mertie wrote: "Panning reveals the nature and amounts of accessory minerals, which are of high significance and can be correlated with the petrology."

In his study of granitic rocks in five southeastern states, Mertie panned 55,415 pounds of decomposed or crushed bedrock to identify a belt of monazite 620 miles long. Each sample panned averaged eighty-two pounds. When you consider that a heaped sixteen-inch pan held about twenty-two pounds of material, and that each pan was repanned about four times to capture the semi-heavies, the amount of time he spent panning each sample works out to two hours!

While anyone can recover nearly all the coarse gold from a placer sample in a few minutes, retaining semi-heavy minerals and gold that approaches flood-gold fineness requires much more time and technique. Mertie felt thirty minutes for panning one heaped pan was not unreasonable—and that's just the initial panning in a multi-pan process.

Although most people are not panning eluvial or decomposed bedrock material, this may occasionally be necessary when closing in on a lode source. In general, the use of Mertie's panning techniques will be more useful when evaluating stream drainages for economic minerals. Rather than a quick panning operation for gold, today's prospector has to be open to the wide spectrum of deposits that may be found in the concentrates of a gold pan.

Because of the back-breaking nature of panning, Mertie felt that fourteen pans or seven hours of work constituted a good day's effort. Since panning is done in about six inches of water in a quiet eddy, rubber boots are needed, and some of the discomfort of bending over can be eased by sitting on a small stool or overturned bucket .

Three or four pans are employed, in addition to screens for rejecting the oversize gravels. One pan is used for the original material. The second pan receives outwash material when the first pan is two-thirds empty. The last and heaviest concentrates from the first pan go into a third pan. The first pan is charged with another load of creek material and after two-thirds is rejected, most of the final one-third goes into the second pan, and the heaviest concentrates go into the third pan. After three or four pans are panned in this manner, the large fraction in the second pan is repanned, reserving most of the last one-third for possible repanning while the purest concentrates are added to the third pan. The third pan then has a concentrate containing medium to high specific gravity minerals.

Cleaning of these concentrates to eliminate silicates such as quartz, epidote and garnet is the next step in the process. Mertie describes this as "washing, floating, and wiping." Crystals and mineral particles coarser than sand are caused to walk or roll out of the pan by the in-and-out flow of water. A collecting pan is used to receive the rejects so that semi-heavy losses can be recovered from this material by further panning.

Vigorous agitating causes the fine quartz and similar low density minerals to rise in suspension. By quick action, the water containing the particles can be poured off into a collecting pan for repanning. If some of the suspended particles sink before reaching the lip of the pan, these are wiped out with the thumb. A lot of patience is necessary in these last stages of the process to eliminate quartz and other rock-forming minerals. Panning in sunshine helps show that minute particles are being eliminated even when this is not otherwise apparent.

"How do you know you are not washing out micro-gold?" Professor Madonna once asked me after we had discussed the pan concentrating process. Unfortunately, the answer is that fine gold is being washed away at every step of the procedure. This problem was addressed by Milt Wiltse in a paper for the Alaska Division of Geological and Geophysical Surveys called "The Use of Geochemical Stream-drainage Samples for Detecting Bulk-Minable Gold Deposits in Alaska."

Wiltse noted, "Conventional methods of obtaining heavy-mineral concentrate samples produce results strongly biased toward coarse-grained gold...[and] they are not always useful for detecting low grade bulk-minable deposits"(p.3). Wiltse felt that panning as used today is not very reliable, because of small sample size (a single pan will not detect subtle anomalies), hydraulic variance between streams and sample locations within a stream, and loss of fine-grained material having micro-gold. He concluded that "...retaining the sample's fine-grained size fractions is fundamental to effective geochemical drainage surveys for bulk-minable deposits" (p. 8).

Finding a balance between retaining heavies and semi-heavies in a concentrate is the critical point. One method is to assay the concentrate from a large sample–three or four pans from one location–to identify the elements present. Or, the two fractions of heavies and semi-heavies may be separated out by panning and mineralogical identification made through magnets and microscope. Separating the two fractions involves more panning with multiple pans. Initially, a large part of the heavies and semi-heavies is washed into a second pan in order to retain a sample free of semi-heavies in the first pan. Panning of the second pan into a third pan will recover most of the heavies. Mertie repeated this process up to ten times to make a complete separation of the heavies and semi-heavies.

The concentrates are then air dried and the heavies are separated by various strength magnets. Magnetite is drawn off with a weak magnet, then ilmenite and pyrrhotite and other minerals having a similar magnetic susceptibility are separated out, and in the last step a very strong magnet is used for the weakly magnetic fraction. The leftovers contain non-magnetic minerals such as zircon, tungsten, cassiterite, and pyrite. This last fraction and the semi-heavies are examined with a ultraviolet light and under a microscope, for crystal structure, color, and other identifying characteristics. Semi-heavies could contain corundum, zircon, rutile, tourmaline, garnet, precious and semi-precious gems, and diamonds.

Paul Theobald also examined the gold pan's ability to capture minerals in, "The Gold Pan as a Quantitative Geologic Tool." Factors which he identified as having the greatest effect on recoveries are the specific gravity of the minerals, their grain size, and the proportions of the various grain sizes. In a study of 4,000 samples, recoveries were best for elongate minerals like zircon, rutile, and epidote; these semi-heavy minerals resisted the buoyant effects of water during panning, and they rolled in only one direction.

He found that equidimensional minerals such as garnet and monazite exposed a smaller area for the buoyant effects of water, but they rolled in any direction and were likely to be lost. Large particles also presented a greater surface area, which absorbed more force from the water moving out of the pan. Care must be taken in the early stages of panning because many of the coarser grains may be rolled out of the pan. Although plate-like minerals such as muscovite and biotite were abundant in the stream sediments, neither was recovered in the concentrates.

Theobald concluded that minerals with specific gravity of 4.5 or greater had the best recovery. Even with a specific gravity of 3.2, minerals could be recovered, although at lower percentages. By retaining a large proportion of quartz (specific gravity 2.6) in the concentrate, the amount of minerals whose specific gravity is around 3.5 is increased. This means that the gold pan would capture more epidote, diamond, realgar, garnets, chrysoberyl, sphalerite, zircon, chalcopyrite, and other minerals in the specific gravity range of 3.4 to 4.3. However, in spite of careful, multiple stage panning, a few grains of heavy minerals were invariably lost.

If it is abundant, fine-grained quartz will accumulate in the bottom of the gold pan and prevent settling of the heavier minerals. In addition, fine-grained heavy minerals can be lifted into suspension nearly as readily as quartz. Theobald recommended that in the later stages of panning the concentrate be passed through a sixty-five-mesh screen (screen with sixty-five holes per linear inch) and the quartz panned from the undersize. This prevents the loss of fine heavies with a specific gravity matching larger sized quartz, and answers Wiltse's concern for retaining fine-grained material for an effective analysis.

The good news is that few of the twenty-six panners in Theobald's study were familiar with the gold pan, yet recoveries were high from the beginning of the experiment. There were no systematic increases in recovery with increased experience and no differences between panners. This means that anyone can achieve professional results following the screening and multiple panning techniques outlined by Mertie and Theobald.

SUMMARY

Through the use of multiple pans, careful panning will save a greater portion of low specific gravity minerals that provide clues to lode deposits in the drainage. The potential for finding precious and semiprecious gems, pathfinder and indicator minerals, and diagnostic rock-forming minerals is greatly increased. Finer and lighter-weight gold will be captured. Since most streams have only been panned for a few distinctive heavies, use of the gold pan to analyze for semi-heavies provides a new method for finding mineral deposits.

REFERENCES

Mertie, J. B. 1954. "The Gold Pan: A Neglected Geological Tool." *Economic Geology, Volume. 49*, p.642.

Mertie, J. B. 1979. "Monazite in the Granitic Rocks of the Southeastern Atlantic States–An Example of the Use of Heavy Minerals in Geologic Exploration," USGS *Professional Paper 1094.*

Theobald, Paul. 1957. "The Gold Pan as a Quantitative Geologic Tool," USGS *Bulletin 1071A.*

Wiltse, Milt. 1988. "The Use of Geochemical Stream-drainage Samples for Detecting Bulk-Minable Gold Deposits in Alaska," DGGS *Report on Investigations 88-13.*

7. Before the Klondike: Circle, Alaska

A large excavator was clanking off a lowboy trailer when I arrived at the Circle Mining District Museum for the annual picnic and rodeo. Taking a break during the busy mining season, placer miners from this historic district 130 miles northeast of Fairbanks, Alaska, get together for a day of socializing, fund-raising, and games. It's a volunteer effort to raise money for the district museum and the miner's association.

MINING HISTORY OF THE AREA

Shortly before Alaska's purchase from Russia in 1867, Father Robert MacDonald, an itinerant minister for the Church of England at Fort Yukon, had spread stories of a gold-rich creek located somewhere south of the Yukon River. Prospectors searched for "Preacher Creek" in the Yukon-Tanana Uplands, an area between the Yukon and Tanana Rivers in Interior Alaska where new gold deposits are still being discovered. The first hint of gold came in the 1880s when American prospectors found placer gold near the Alaska-Canadian boundary on the Forty Mile River. Jack McQuesten is called the "Father of the Yukon" because he grubstaked many prospectors, and built an Alaska Commercial Company store at the mouth of the river.

Two hundred miles down the Yukon from Forty Mile and across thirty miles of swamp and spruce flats, men found gold on Birch Creek in 1893. McQuesten set up a store at nearby Circle City. Named Circle because people thought the Arctic Circle passed through the area ("a clearly marked line in the forest," sourdoughs, slang for someone who has overwintered in the north, joked to newcomers), it became the largest log cabin city in the world. Miners built small homes in the town and made the trek to the diggings along creeks that drained a distant ridge of low hills.

Although it brought prospectors to the area, Birch Creek never became important as a placer creek. However, since many of its tributaries were gold-rich, the placer mining operations on them fed silt into Birch Creek and eventually caused public concern that led to strict environmental regulations. Now part of the White Mountain National Recreation Area and a wild and scenic river, Birch Creek is closed to mining and is a prime grayling habitat. On the upper 140 miles of the creek, canoeists can also pan small quantities of gold almost anywhere.

PRESENT DAY PLACER MINING

"I started visiting mines around here in the summer of 1991," Kathy Charlie, an Alaska Division of Mining employee, told me. "I got very familiar with peoples' operations because I was helping them with their plans. Permits are getting harder and harder. It was one page and now it's ten pages."

Miners stopped by to greet her and ask when she would be coming out to their creek for a site visit.

"In the early '80s, people didn't worry about settling ponds," she explained. "Then miners made the transition to ponds and a one hundred percent recycle. As long as the gold price was high, the cost could be absorbed."

"Reclamation wasn't an issue at first, but in 1991 the state began reclamation requirements. If it's done right, in five years you end up with better moose browse, young willows and duck ponds. A nice eco-niche. But it upped the cost of doing business. These miners have hundreds of thousands of dollars invested in equipment. It has to stay simple and reasonable [to be economically feasible], but more and more is being required, to a point of diminishing returns. If there's a five-year low gold price, most of these guys will have to move on."

Circle proved that paying quantities of gold could be found in Interior Alaska. Over one million ounces was taken from Mastodon, Mammoth, Deadwood, Harrison, Ketchem, Crooked, and other creeks. The Circle District was a placer area open to pick-and-shovel miners. In a ten-hour day, one man shoveling gravel into a sluice box could recover two to three ounces of gold.

The history of this district illustrates a hundred years of change in placer mining technology. Many of the creeks have been profitably mined five and six times, utilizing different equipment and working lower paying areas.

"Me and the family first mined on Frying Pan Creek," Bill Ohman told me. "It was frozen ground. My equipment was too small. The economics of it was against me. Then I got a partner, and we made a living on Squaw Creek for four or five years. We had big machinery: a 988 loader, large excavator, and a mobile screen plant. We moved a lot of gravel. On Harrison Creek we mined by jumping around on where it'd been mined before. We'd find places between cuts, under tailings, and on the side of the creek."

A 1991 USGS study of the Circle gold placers (*Bulletin 1943*) characterized some of the creeks as "mined out," (Yeend, p.12) and "thoroughly mined" (Yeend, p.14). But even though they've been almost continuously worked since their discovery, Circle's creeks still support a few active placer operations.

"There were over a hundred and forty operations in this district in the early '80s," big Ed Lapp recalled. "Now you can count them on two hands, maybe on one hand."

"This is our fourth year on Ketchem. The boys were in here the tenth of April, and we'll try to end up the fifteenth or twentieth of September. There's four grandsons, my son, Clayton, and his wife, Joan. My wife takes over when Joan leaves.

"We started small, rented our equipment. Then we rented a CAT and bought a small loader. We ran a summer operation on Eagle Creek for quite a few years.

"Ever since we started in, it's been a family operation. Before the kids go back to school, we keep two shifts going twenty hours a day. You've only got so many days. It takes $150,000 a year to run this place. Last season we got 1,600 ounces. Clayton runs the D-9 high track, and my grandson, Boman, operates the excavator." Apparently Boman's experience paid off...at the picnic anyway.

On Ketchem Creek, I watched Clayton Lapp in the D-9 CAT pushing up gravels for his son to feed into the wash plant. Boman, encased in the cab of the excavator, dumped a bucket of material into the top of the plant, slowly in order to prevent clogging or plugging up the flow. With precision, he scooped up bucket after bucket of gold-bearing gravels and fed them into the sluice.

EARLY PLACER MINING

"In the past we had an eleven-year-old win on the excavator," George Hiller, president of the Circle Historical Society, informed me at the picnic. "Family operations teach their children. By the time they're grown, they're experts at using and fixing the equipment. You gotta have the concentration and diligence to sit there all day and focus...as boring as it might be."

In the early days, shallow open cuts and shafts through the permafrost ground exposed the richest gravels. Men would haul wood and build fires on the ground to melt the permafrost eight inches at a time. Ashes, mud, and gravels were shoveled or winched out and a new fire built, until bedrock was reached. Wood fires were soon replaced by small steam boilers using hot water to thaw the gravels.

Interior Alaska valleys had been covered with Pleistocene debris and silt which had frozen, blanketing the ancient creek channels. Modern creeks had no relation to the old channels, so miners dug shafts across the valleys to locate the real paystreak. Once the shaft struck pay, miners tunneled underground along the channel that followed the richest area. Winter was best since frozen water wouldn't fill the tunnels, though when the men emerged from the steamy workings, their muddy clothes would freeze. Pneumonia and tuberculosis from the miserable working conditions and poor diet often led to death.

Circle was the supply and social center, the "Paris of the North," where miners came to get supplies and met to decide the size of the claims and settle disputes. If necessary, a judge and jury would be appointed and a case tried in a few hours.

Twenty-eight saloons and eight dance halls enlivened the town. Jack London set the beginning of his novel *Burning Daylight* in a Circle saloon.

"Now supposing, Wilkins," Daylight went on, "supposing you all knew it was going to rain soup. What'd y'all do? Buy spoons, of course. Well, I'm buying spoons."

Based on a real person, Daylight made his fortune buying mining claims on Bonanza Creek in the Klondike and in the Forty Mile and the Circle Districts. Buying, selling and leasing claims actively continues in the Circle Mining District.

Circle's new school only lasted one winter before the fabulous Klondike Strike emptied the town. The Tivoli Theater and the Grand Opera House lost their customers to Dawson. By late spring of 1897, only fifty people, mostly women and children, were left out of what had been a thousand.

Within a year, however, miners returned to Circle after finding the Klondike creeks already claimed and restaked their claims. Soon, gold was discovered on Mastodon Creek, and for a while it was the richest creek in Alaska. Upstream tributaries of Mastodon were Mammoth and Independence, and downstream it became Crooked Creek and all had gold-bearing placers. Gold was also found on Deadwood (at first called Hog 'em Creek because the discoverer filed claims for family and fictitious friends to cover the ground), Ketchem, Harrison, and Porcupine Creeks.

As the district developed, wooden sluices were replaced by steel ones. The addition of a steel sluice plate at the front of the sluice eliminated the need for a bedrock drain and allowed an easier feed. Portable dredges were tried on some of the creeks, but hydraulic mining was the most popular method after 1909. Long ditches brought water from upstream sources and delivered it from a high point to generate enough water pressure for the hydraulic giant or nozzle to break up the ground with a water jet.

In the 1920s, steam-powered draglines, cable excavators, and scrapers were used with elevated sluices to handle greater volumes of gravel. With an elevated sluice, the tailings coming off the end of the sluice box didn't have to be constantly removed and stacked.

Then, more mobile diesel-powered bulldozers, draglines, and efficient water pumps revolutionized open-cut placer mining. The increased mobility made it possible to work deposits that had been untouched by the more cumbersome equipment. By the 1980s, D-9 CATs, excavators, and front end loaders worked benches and lower paying valley margins. More sophisticated screening and better wash plants captured finer gold.

MODERN PLACER MINING

Ed Lapp's operation on Ketchem Creek now runs about one hundred thirty cubic yards of gravel an hour.

"We screen it down to half-inch," Lapp informed me. "We get very few nuggets on this creek. We have a two-deck shaker screen and a six-foot wide sluice with 'live' riffles in it—water injected, to keep the riffles clean. Fine gold will stay right up there. Some of it's just like powder."

Ed Gelvin smiled when he remembered selling his gold to the U.S. Mint for $30 an ounce. "We mined on Greenhorn, Squaw, and Birch Creek over by Gold Dust and on Crooked Creek. The whole family helped out one way or another. I had a D-7 CAT and just a sluice box on Squaw Creek in 1954. Later, we had D-9s, loaders, backhoes—all kinds of equipment. When I retired, my son mined on Ketchem Creek."

A modern placer mining operation near Central, Alaska illustrates the equipment needed. This family operation works two ten-hour shifts a day.

A crude road was eventually built from 1928 to 1932 along the stampede trail that led to Fairbanks. People could then drive out in the summer and set up camp. From Fairbanks, the 128-mile well-maintained dirt and gravel road to the mining area follows the Chatanika River to its headwaters, winds up Twelve Mile Summit, (the divide between the Yukon and Tanana River drainages) and passes over Eagle Summit before running downhill above Mastodon and Mammoth Creeks and into the black spruce plain around the town of Central. Between Twelve Mile Summit and Eagle Summit, the main access point to Birch Creek gives canoeists entry to a long stretch of this wild and scenic river.

Visitors seldom go to Circle, (thirty-three miles beyond the placer creeks) unless they wish to wash their feet in the Yukon, launch a riverboat, or visit the old pioneer cemetery, (a ten-minute walk through mosquito-infested lowlands). Circle's population of around eighty-six people supports a trading post and gas pump.

Central, thirty-two miles west of Circle, draws more visitors than Circle because of its location at the crossroads to a nearby hot springs; its log cabin museum and school; and Crabb's Corner, a friendly restaurant and store with gas, motel, and camp sites. Set among birch and spruce trees, and close to the mining areas and the hot springs, many people make Central their summer home.

"There aren't any hook-ups for RVs," Jim Crabb admitted. "It's for camp camping–dry camping with an RV or tent. Our log home is there. We love to sit on our porch in the evening and visit with campers. People who come out this far are really interested in the area."

Dog mushers come through here in February during the grueling 1,000-mile Yukon Quest Sled Dog Race from Whitehorse to Fairbanks, or the reverse (in alternate years).

"Dog mushers get a free steak meal," Crabb explained. "They stop over for ten to eighteen hours, resting up before going up Eagle Summit or after coming off of it."

Near the town's main street is the Circle District Museum, with its display of gold from the different creeks and collection of artifacts from the old cabins in the area. Loose-leaf binders of photographs show the old dredges and historic mining in the area. In addition, the museum archives has interviews with pioneers and a research library.

From Central a road leads eight miles east to Circle Hot Springs a historic sulfur hot springs with a swimming pool and hotel-restaurant, where miners have soothed and soaked themselves since they first came here. On the way to the hot springs, the road crosses Deadwood and Ketchem Creeks, and dirt roads leading up these well-worked valleys make a nice side trip to view placer-mined valleys. Granite tors near Ketchem Creek provide a picnic site with a panoramic view. Beyond the springs, the road continues on to Portage, Bottom Dollar, Half Dollar, and Harrison Creeks, where placer activity also continues.

SUMMARY

The origin of the placer gold in the district's creeks has not yet been discovered. Two elongate bodies of granite form the low hills, but creeks draining them contain cassiterite and wolframite, making them tin granites. A dark schist seems to hold the clue since where that appears, the creek has placer gold. Mastodon Creek, one of the most productive creeks in the district, is largely within this schist. It's possible that erosion destroyed the original gold lodes, or poor exposures may still mask the source rocks.

Although many of the creeks have been worked out, Circle continues to be best-known as an important placer area. These creeks brought prospectors to the Yukon even before the Klondike was discovered.

8. The Search for Diamonds

When I first saw the shining bluish mass of biotite, I couldn't help picking up the rock for a closer look. Unlike anything I'd ever seen, it had a radiance that set it apart from the quartzite, schist, and granite of the area. I sat down on a ledge overlooking the spruce-covered valley to ponder the origin of this oddity.

"Lamps" my geology teacher later told me. "Short for lamprophyres and lamproites."

This comes from the Greek word for 'glistening' because many of these igneous rocks contain reddish-brown coppery phlogopite mica. Lamprophyres are distant relations of kimberlites. Like an elevator, lamprophyre and kimberlite magmas travel at speeds of around twenty-five miles an hour on their explosive journey to the surface. Composed largely of olivine and potassic- and magnesian-rich minerals from the mantle, they pick up "passenger" rocks which form breccia as they pass through different layers on their way to the surface.

RECENT DIAMOND DISCOVERIES

The diamond world was turned upside down after discoveries in western Australia in the 1970s indicated other sources besides kimberlite pipes for these gemstones. One of the rarest of all rock types, lamprophyres came under intense scrutiny after their link to diamonds was discovered. Some known kimberlites were even reclassified in light of new knowledge about lamprophyres. The Prairie Creek kimberlite at Crater of Diamonds State Park in Arkansas is now known as a lamproite, a diamond-bearing type of lamprophyre. The Yogo sapphire host rock in Montana is also related to a lamprophyre dike.

This clan of potassium- and magnesian-rich rocks has a "daunting and incomprehensible terminology" according a recent text on them. My mind goes numb reading about fitzroite, wolgidite, wyomingite, minette, alvikite, and shoshonite. Though less common than kimberlites, lamproites are far richer in diamonds. The Australian Argyle Deposit has 500 carats per 100 tons as compared to the kimberlite average of 20 to 80 carats per 100 tons. Both rock types are mantle-derived from regions of thick crust, but lamproites appear on or near craton (the stable interior continental plate) margins.

The minerals that are traditionally used to locate kimberlites—pyrope garnets, chrome diopside, and ilmenite—are not always useful for finding lamproites. Sometimes diamonds themselves serve as the best indicators. In 1979, after seven years of searching, geologists traced diamonds and chromite twenty miles up a stream to the Argyle Pipe in the far north of western Australia, where a linear lamproite measuring almost 5,000 feet long had eroded to form a narrow valley. While the rarity of diamonds does not make them an ideal target to follow, when they are found, they can lead to their original source.

Based on the size of the diamonds, gravels three-fourths- to one-sixteenths-inch were processed from the alluvial deposits in two streams that drain the Argyle deposit from 1983 to 1985. Then, an open pit on the lengthy lamproite dike was developed in 1985.

Hardrock material is crushed, scrubbed, screened, and processed through several kinds of gravity separation. The final concentrate was passed through x-ray sorters, which cause the diamonds to fluoresce, making them easier to see and collect. The average size of Argyle diamonds is less than one-tenth of a carat, relatively small compared to kimberlite diamonds. In addition, only five percent of the Argyle pipe's diamonds are gemstone quality and an additional forty percent are near-gem quality, low by world standards. Even so, the mine has been very profitable.

DIAMOND PROSPECTING

Diamond prospecting begins with washing and classifying or sieving stream gravel or eluvial material through various sized screens. So mining companies send out geologists to collect large samples of stream gravels or soils, which are washed and screened in the field to reduce their bulk before they are brought back to the laboratory for additional processing and visual analysis. "Diamond Exploration Techniques Emphasizing Indicator Mineral Geochemistry and Canadian Examples" by C.E. Fipke, J.J. Gurney, and R.O. Moore (Geological Survey of Canada *Bulletin 423,* 1995) explains the types of kimberlite and lamproite minerals, and provides a good list of references.

In a *Rock and Gem* magazine article (April 1999) James and Jeannette Monaco, authors of *Fee Mining and Rockhounding Adventures in the West* and *Fee Mining and Rockhounding Adventures in the East*, described how James Archer, an expert diamond finder, conducted his search for diamonds in Crater of Diamonds State Park in Murfreesboro, Arkansas. They spent a week with Archer, washing and screening gravels and learning how to use a "saruca," (Monaco, p.33) a screen with a slightly concave center to stratify the gravels. During the washing and shaking, heavier specific gravity minerals settle to the bottom of the screen. Then, by quickly turning the contents over on to a prepared surface, the heavier minerals end up on top of the pile where they can be easily seen.

Small-time prospectors use gold pans to collect and wash down the screened material in their search for indicator minerals. With a specific gravity of around 3.3, about the same as quartz, diamonds don't settle out as easily as cassiterite or gold, but the more numerous pathfinder garnets do show up. I always feel I've done a successful panning job if garnets show up in the bottom of my pan because diamond prospecting is all about garnets. Normally a person doesn't have diamond-bearing gravels and kimberlite to start with, but looks for the pathfinder minerals such as pyrope garnet, chromite, chrome diopside, and non-magnetic magnesian ilmenite, a titanium mineral.

I visited one of our local jewelers to compare the purplish color of pyrope garnets with the reddish, brown, and orange varieties. Taylor's Gold-n-Stones in Fairbanks, Alaska displays a variety of cut and raw garnets.

"We get pyrope garnet from the Navajo reservation in Apache County, Arizona," a smiling Glenn Taylor told me. "We used to live there, and our Dad still operates a trading post. The Navajos put pyrope in gourds for rattles because they last forever."

The Taylor brothers also dredged gemmy, deep red almandine garnet from Alaska's Forty Mile area as part of a gold mining venture in the early 1980s. Gem-grade faceting garnets are only found in a few Alaskan streams, but when set into or beside gold nuggets they add fire.

"We buy brownish-red almandine garnets as big as marbles from Wrangell in southeast Alaska, but they are basically sandpaper material," Taylor conceded. "These Wrangell garnets are found in the schist."

A knowledge of the types, colors, and origins of garnets is essential when checking gold pan concentrates for diamond indicators. Purple to pink pyrope garnets are potential kimberlite indicators. Chrome diopside, another important kimberlite mineral, has a pale- to medium-green color, but since it doesn't travel far from the kimberlite pipe before disintegrating, it only shows up when the search is within a mile of the source. Chromite and ilmenite are found in a variety of settings, so they aren't reliable minerals to look for unless pyrope garnets are present too.

CHARLES FIPKE

With the new lamproite type of diamond target, the search for diamonds has been greatly expanded. Another great modern diamond success story centers on Charles Fipke, a Canadian geologist, who was hired to look for kimberlite pipes in Colorado's Rocky Mountains in 1978. The company moved him to the Northwest Territories' Mackenzie Mountains where Fipke continued to search. Even after the company ceased its diamond project, Fipke was convinced he was on to the trail. Unfortunately, following up pathfinder minerals takes perseverance because ninety percent of kimberlite pipes don't have diamonds.

Fipke formed a partnership with another geologist, and they worked their way east toward the Mackenzie River, retracing the path of a Pleistocene glacier, which had brought the garnets, chromite, and ilmenite in its ice train. By 1985 they were out of money, so Fipke formed Dia Met Minerals, and sold shares to raise capital. All summer he flew out collecting samples, and while analyzing them that winter, he found one that had 6,000 pyrope garnets!

Camp was set up in 1989 on Little Exeter Lake, two hundred miles northwest of Yellowknife. When asked, Fipke told people he was prospecting for gold. Out of time and at the end of his budget again, he investigated a circular lake that suggested the end of a tubular pipe. On the lake shore his son found bright green chrome diopside, confirming they were on top of the pipe.

Fipke quickly negotiated an agreement with Broken Hill Proprietary (BHP), an Australian mining conglomerate, and BHP poured millions into intense exploration. A drill hole angling under the circular lake hit kimberlite, and BHP announced in 1991 that they had found eighty-one small diamonds in the hole. The news thrust the Northwest Territories into a mad staking frenzy, and since then more than hundred pipes have been discovered on the BHP claim block alone, with about a quarter of them being diamondiferous. BHP has sampled twenty-two pipes on its claims for diamonds, and four are now in production. Vernon Frolick tells Fipke's story in *Fire Into Ice: Charles Fipke and the Great Diamond Hunt* (Raincoast Books, 8680 Cambie Street, Vancouver, British Columbia, Canada V6P 6M9).

DRILLING DIAMOND PIPES

Frank Smith (not his real name), a young geologist with an engaging smile and youthful enthusiasm, joined Canamera Diamond Company and later BHP, searching Canada's far north for diamond pipes. While on an Alaskan project, he told me about his experiences.

"There were four camps with twelve to fifteen people in each," he recounted. "We were covering a vast area. The company would fly us out each day. We picked out our traverses to avoid lakes and marshes. Some places were boulder fields, good ankle twisting country. It was fun. I liked hiking around in the loon dung (Canadian slang for a very muddy area with mosquitoes and black flies in summer that is also underlain by permafrost so it is poorly drained).

"We used a shovel, a ten-mesh screen [ten holes to the inch] to take out the big stuff, and a large pan that didn't fit in the pack. Even the clays were decanted and saved. We got our best samples from frost heaves where the soil was pushed up. We also got samples by the sides of lakes and in streams. We'd bag ten to twelve samples a day covering six miles, and a helicopter would come by later to pick up the flagged bags.

"The material was sent to a Vancouver lab where they screened it, put it through a density separator and drew out the magnetic portion. The remainder was dried and examined under a microscope for pyrope garnets, chrome diopside, and magnesian ilmenite. I've seen thousands of white plastic buckets of samples outside of those labs waiting to be analyzed."

Because of the volume of material, the company began doing their lab work on site. The heavies and the semi-heavies were removed using cyclones.

"Once we got rid of the olivines, it was easier to see diamonds," he recalled. "I saw a diamond on the shaker table once—about ten carats. It was bouncing around, so I picked it up. They wouldn't allow that now, of course. It was rectangular, about the size of my pinky fingernail. You could see the diamond facets under a microscope."

He worked during the winter on a drill rig over one of the pipes on a frozen lake.

"We pin-cushioned the whole pipe," he said. "It weakened the ice, so we'd move the drill rig to the other end of the lake after a few holes to give the area a chance to freeze up again. We stacked tons of sample bags filled with drill chips next to the rig."

Pipes were invariably located under lakes since the kimberlite eroded faster than the surrounding rock. Once a likely area was identified, airborne magnetic surveys could pinpoint pipes because of their magnetic minerals. Since pipes are usually only two and one-half to ten acres in size, their magnetic signature was often the best way to find targets quickly. Pipes also seem to occur in clusters of six to forty, causing companies to stake large claim blocks.

"If it hadn't been eroded or scraped off by glaciers," Smith told me, "the crater would contain [volcanic] ash tuff, and stream gravels that had slumped in after the glaciers melted. I've seen perfectly preserved pieces of wood from the Jurassic that fell into the crater after it was formed.

"Below the crater facies the drill encountered tufaceous kimberlite in the volcanic neck while hydrothermal fluids and calcite created a gummy substance that made drilling difficult. The garnets and olivine had white calafite rims from carbonate alteration [during their gaseous and magmatic transport to the surface]. Some garnets would be so altered that there would be only a tiny purple core surrounded by the carbonate.

"Drilling got harder as I got into the deeper hypabyssal area. Magnetite increased, but there were less pyrope and more orange garnets derived from the high grade metamorphic rocks. The matrix was hard and black with a lot of olivine."

Kimberlite from deeper portions of a pipe is a dark, brecciated mass of resistant rock, which South African miners called "hardebank" because it doesn't break up.

Kimberlite pipes usually form at seventy or more miles depth beneath thick, stable cratons such as those found in South Africa, Siberia and the Precambrian Shield of Canada's north. The "cool roots" of thick continental cores seem to have the right mix of high pressures and temperatures necessary for creating diamonds and for transporting them intact in kimberlite pipes.

Production began at BHP's Ekati Mine, Canada's first diamond mine, in 1998. Just sixty miles south of the Arctic Circle, it has access to a winter ice road to Yellowknife, one hundred eighty-five miles southwest. Since then the Panda, Koala, Sable, Fox, and Misery Pipes have been developed, and they will supply material to a centrally-located processing plant for at least seventeen years. Conventional open pit mining methods will be used on many of the pipes, but the Panda and Koala Pipes have sufficient values to justify underground mining. The Ekati Mine expects to produce three and a half million to four million carats annually.

Charles Fipke owns ten percent of the mine, and his Dia Met Company has another twenty-nine percent. This determined geologist who had a vision is now worth about $350 million.

ALASKA DIAMONDS

I heard about a 1980s diamond find in Interior Alaska. Placer miners 120 miles northeast of Fairbanks, in the historic Circle Mining District, had discovered three diamonds in their sluices. Since equipment used to recover alluvial gold is set up to capture very high specific gravity minerals (gold has a specific gravity of 15 to 19), lighter materials such as diamonds and their indicators, (specific gravity 3.2 to 3.5) would wash right through a gold sluice. So gold-bearing gravels that have given up their gold may still contain diamonds. How many diamonds have been washed back into these streams, I wondered.

Ranging from one-third carat to one and one-half carats, the Circle diamonds were light yellow, yellow-white, and clear. They had dodecahedron and octahedron shapes and were rounded and fractured, which suggested that they had been moved a long distance from their source.

The DeBeers Group, a diamond cartel that controls the sale of diamonds, sent representatives to the Circle area to investigate, and some State of Alaska geologists tested the gravels for indicator minerals. However, the gravels were deep, there was no specialized diamond processing plant available, and the geologists didn't find any diamond-indicator minerals. Their report concluded that the source could be in any of several terranes in Interior Alaska.

I had been working about forty miles away from the diamond discoveries in a gold placer area, patiently identifying the different rock types that could have created the gold. The shiny, coppery-colored lamprophyre I came across soon led me to do more detailed prospecting in the area. Some prospectors feel lamprophyres are a good indicator of lode gold deposits, since they are often found in close proximity. I did notice small gold anomalies showing up in my soil assays whenever there was an outcropping lamprophyre uphill.

I dug into the hillside and pulled out reddish oxidized samples of the lamprophyre. Cracking them open, I found large, red-brown garnets. Unfortunately, garnets in kimberlites and lamprophyres may come from a variety of lower and upper mantle sources, and these may be incorporated into the lamprophyre magma as it moves toward the surface. Since only the chrome-rich pyrope garnets originate from deep sources, the "diamond stability field," these garnets are considered the best indicators of deep mantle material and diamonds.

Do I have a potential diamond source? Probably not. Assays of the rocks don't show a high magnesium or titanium content in the lamprophyre, important indicators of lower mantle origin. In order to rule out any chance of diamonds, however, I will need to process a great deal of the creek gravel. This involves screening out the larger-sized material, thoroughly washing the gravels to remove clay coatings, and then using gravity separation with jigs and cyclones to focus on the lighter specific gravity diamonds. A grease plate could be used as a trap. Examination of the concentrates under a microscope and an ultraviolet lamp would complete the process.

SUMMARY

Lamprophyres do not have the distinctive pyrope garnet, ilmenite, and chrome diopside indicators that are used in the search for kimberlites. Placer diamonds downstream of lamproites remain the best pathfinders. A high magnesium and titanium content of the rocks is also important. Large garnets in the matrix may be useful, but only if they are chrome enriched. Since diamond-bearing lamproites have only recently been recognized, new studies and additional finds will undoubtedly increase our understanding of them.

Because of their unique genesis in the mantle under thick continental rocks, diamonds and their host rocks provide a fascinating window on deep crustal minerals and processes. Kimberlite was first recognized in South Africa in 1870. New discoveries in Australia have expanded the search for diamonds to include lamproites. Prospecting techniques are still adjusting to these "glistening" rock types and their paucity of indicator minerals.

9. Nevada Monsters: Carlin's Microscopic Gold

A motley group of geologists struggled up the steep, sage-covered slope to a prominent outcrop of dark conglomerate. Below and surrounding the group were the massive open pit copper workings of Copper Canyon and Copper Basin, and the Fortitude and Midas Gold and Silver Mines. Ted Theodore and Jeff Doebrich led the group. Their detailed USGS geologic map of this Battle Mountain area of northcentral Nevada had just been published, and they knew every fault, rock type, and geologic formation of the area. Theodore stood on higher ground and addressed the group.

"It was here that Ralph Roberts first understood the overthrust that led to the Carlin discoveries," he explained. "Roberts saw the differences between the sedimentary rocks of the Battle Formation and a plate of early Paleozoic rocks that had overthrust the area–an event he named the Antler Orogeny."

One arm swept across the western horizon. "Although haul roads and waste dumps have eaten into this outcrop, I hope that it will be preserved as a landmark to his accomplishments." A few geologists approached the rocky buttress behind him with their pointed rocks hammers. "This is an example of geology at its finest," Theodore continued. "The discovery was enormously exhilarating to Roberts. In the 1930s and 1940s he was a junior to most of the USGS geologists who were working in Nevada. Together, these geologists established a regional tectonic framework for northern Nevada that led to the understanding of the gold deposits in the Carlin Trend."

Later that day, near the floor of the Fortitude Pit, the location of the Golconda Thrust was pointed out. This was where the plate had come in from the west and overran limestones and other shallow marine sediments.

The Battle Mountain area of north-central Nevada contains copper and gold mines. The open pit Fortitude Gold Mine is close to where Ralph Roberts, USGS geologist, unraveled the geology of the Carlin Trend.

"If you can find the contact, you've got better eyes than I," admitted Pat Wotruba, one of Battle Mountain Company's geologists. "I'm used to seeing a big crush zone and stuff smashed up. Here, there's almost nothing to indicate the overthrust."

"The impermeable rocks carried in by the Golconda plate served as a cap to ascending mineralized fluids," Theodore pointed out. "The lower plate limestones were reactive and permeable, and that's where most of the mineral deposits were localized."

Roberts described in a 1966 paper (*Nevada Bureau of Mines Report 13*) how this belt extended north through Nevada in Paleozoic time. Overthrusting occurred along the hundred-mile long edge of the ancient western United States' continent, culminating in the eastward movement of great siliceous (quartzite) plates of the Roberts Mountain thrust over the in-place carbonates. The rafted-in plates cover much of northwestern Nevada, but erosion created "windows" (Roberts, p. 68) through the overlying plate, exposing the lower, mineralized rocks.

Roberts wrote, "... [M]ineral belts apparently followed fracture zones that extend deep into the crust. These fracture zones formed the 'plumbing system' that permitted the igneous rocks and accompanying ore-forming solutions to move upward into higher crustal levels. The role of the windows or eroded domes in areas that were covered by the Roberts Mountains thrust plate is significant in localization of ore deposits" (Roberts, p. 68).

MINING ACTIVITY

Since that windy overlook lecture about Ralph Roberts' discovery, I've visited countless mines around Elko, Carlin, and Battle Mountain. Nevada has become North America's richest gold producing area. Ira Joralemon, in his book, *Romantic Copper, Its Lure and Lore*, wrote of "fashions" (Joralemon, p.198-278) in copper mines as they changed from high grade to larger tonnage, low grade operations. Similarly, gold placer deposits gave way to underground gold mines working rich veins. Then open pit mines with disseminated, easily processed oxidized ores became the standard in the gold industry. Now these are giving way to even deeper operations that tunnel into sulfide ores with gold that are below the zone of oxidation.

New techniques have been developed to cope with these refractory ores from which it is difficult to extract the precious metals. Fleets of trucks carry the ore to special mills for treatment. At one of these, Barrick's Goldstrike Mines north of Carlin, Nevada, autoclaves heat the sulfide ore to 420°F under 420° psi pressure, which oxidizes the sulfides, such as pyrites, making the ore more porous and increasing gold recovery from thirty to ninety percent.

Newmont Mining has developed bioleaching for sulfide ores. Bacteria are put into the heap leach pad to break down the sulfide minerals. Newmont then roasts the ore to 1,020°F to burn off the carbon and sulfur pyrites, and recovers the gold. The process produces sulfuric acid as a by–product, which is sold to other Carlin Trend Mining Companies for their autoclave operations.

Oxidized ores are often found in crumbly iron-stained beds, or in impure limestones that have been decalcified into a porous and sugary matrix. Other strata are composed of jasperoid, which was formed as silica was deposited in an ancient hot springs plumbing system. Many rocks have translucent barite veins, native sulfur, and the colorful arsenic sulfides, realgar, and orpiment. Whenever I open a sample bag to examine material I collected from such an area, my hands become coated with reddish dust from the iron oxides.

"Where's the gold?" visitors ask. This microscopic gold in these deposits can't be seen. It's finely disseminated throughout a favorable sedimentary bed. Micro-gold can only be detected through assaying or by the presence of other minerals such as arsenic or cinnabar, the ore of mercury.

THE LANDSCAPE

Almost every spring and fall I join a Geological Society of Nevada weekend trip to visit Nevada mines. Recently, on a typical trip, two buses filled with geologists drove east out of Reno along the "I-80 lineament," a geologists' joke about the Interstate they use so frequently to visit the Battle Mountain and Carlin Trends. A lineament is a geologic term for a major, linear topographic feature of regional extent such as the San Andreas Fault of California.

The interstate east from Reno follows the emigrant trail in reverse. Just east of Reno are the Truckee Meadows where pioneers rested after crossing the Humboldt and Carson Sinks. Now these meadows are being taken over by industrial parks and subdivisions that are spreading out from Sparks, Nevada, where the Pah Rah Range on the north and Virginia Range on the south squeeze the highway and the Truckee River together.

As the river turns north toward Pyramid Lake, the road continues northeast past Fernley Sink. This is the first of several wetlands along the route where migratory waterfowl and shorebirds feed. The Great Basin traps rivers such as the Humboldt and Carson in these low areas with no exit to the sea. Emigrant wagons became mired in the sticky mud and sand of these sinks during their mid-1800s do-or-die-crossing. As we reach Lovelock, the agricultural fields and cows indicate the Humboldt River's freshening away from the alkaline sinks.

The Humboldt River cuts through northern Nevada and is the basis for the ranching economy that started these towns before gold was discovered. The river meanders through low hills, providing an oasis for cattle, birds, and campers, just as it did for emigrants in the past. Mosquitoes can be a problem around the Humboldt in the summer, but clear creeks in the hills are cooler and less pesty.

Geology field trips by the Geological Society of Nevada take participants inside open pit mines in Nevada to view the mineralization. Samples of favorable limestone and alteration can be collected for later study.

In the hot valley between Lovelock and Winnemucca, the Rye Patch Reservoir captures the Humboldt's waters and provides recreation and water for the area. Past Winnemucca, the road crests above Golconda Grade, geologically the area where the Roberts Mountains were thrust inland over the continent's former edge. North of the highway are the Preble, Pinson, Getchall, and Twin Creeks Gold Mines. South, as the road swoops into the flat valley, are the prominent tailings and heaps of the Lone Tree and Marigold Mines.

A small range of hills behind the town of Battle Mountain hides an enormous number of open pit copper and gold mines. This was a center for copper, molybdenum, lead-zinc, and gold-silver deposits, and nearby hills are laced with roads and drill pads. Southeast of Battle Mountain lies the huge gold deposits at Cove/McCoy, Gold Acres and Pipeline, Buckhorn, Cortez, and others. Aside from Battle Mountain these are all discoveries that occurred after Ralph Roberts unraveled the tectonic history of the overthrust plate.

Geologists have returned again and again to these ranges to map the rocks, faults, and subtle alteration of the impure limestones. As deep drilling becomes common, old areas are reassessed, and deposits are found beneath deposits. The drilling penetrates the upper plate rocks, and provides new information about the geology.

CARLIN TREND MINES

Continuing east on Interstate Highway 80 the road climbs Emigrant Pass in the burnt-grass covered Tuscarora Mountains. The town of Carlin sits off the highway and most travelers drive on to Elko's casinos and services. Northwest of Carlin, however, is where this new style of open pit mining really began. With more than twenty deposits containing approximately seventy million ounces of gold, the Carlin Trend may become the largest gold-producing district in North America.

Back in 1907 Fred Lynn discovered small placers near where Newmont's Carlin Gold Mine is located. By World War II, these placers were worked out. Bedrock sources for the gold were identified as uneconomic quartz veins and stringers. These early indicators would not make a mine, given the low price of gold at the time.

In addition, invisible gold that is now being mined from deposits in the Tuscarora Range wouldn't have shown up in gold pans. As W.O. Vanderburg noted in his U.S. Bureau of Mines reports about some Nevada mines in the late 1930s, it was "impossible to distinguish between ore and waste except by assay, and gold is present in such a state that it is impossible to obtain a single color by panning" (Coope, J. Alan. "Carlin Trend Exploration History." *Nevada Bureau of Mines and Geology Special Publication 13.* p. 5). He felt that more deposits would be found in sedimentary formations that tended to have this invisible gold, and he was right.

USGS mapping programs began in Nevada in 1939 and by the early 1950s had extended to central and northeastern Nevada. Ralph Roberts presented his findings in the late 1950s and published in 1960 USGS *Professional Paper 400-B*, entitled "Alinements of Mining Districts in Northcentral Nevada."

In 1961 a Newmont geologist, John Livermore, met with Roberts to discuss these new ideas. With gold at $35 an ounce, bulk-mineable gold properties were still a gamble. However, Livermore convinced Newmont Mining to invest in a prospecting program in Elko and Eureka Counties that would explore the Roberts Mountain's thrust zone for fine-grained gold deposits.

Limited mining northwest of Carlin at the Bootstrap (antimony and gold) and Blue Star (turquoise and gold) Mines indicated that microscopic gold was present, so Newmont got permission to examine the Maggie Prospect, eight miles northwest of Carlin, and got an option to explore the Blue Star. The surface trace of the Roberts Mountain's thrust was readily identifiable, and the tectonic contact was the focus of prospecting. After jasperoid and barite-veined exposures yielded gold values, and further trenching and drilling outlined more gold, a large claim block was staked.

Newmont geologists sampled plain, unmineralized-looking gray float at the Blue Star Mine out of curiosity. This silty limestone was porous from leaching of the carbonate matrix. It assayed a surprising 0.22 of an ounce of gold per ton and led to the identification of the Carlin ore body. The stratabound deposit began production in 1965 and the mine operated for twenty-one years. Since then the mine has gotten new life as deeper sulfide ores have been accessed through underground tunnels below the old pit floor.

Closely following Newmont's development of the Carlin Mine, there were other discoveries along the trend—a fifty-mile long, five-mile wide belt that now includes more than twenty major deposits. It has become one of the premiere gold fields of the world, challenging the Porcupine Mining District of Ontario, Canada as potentially the largest gold producing district in North America. The deposits of the Carlin Trend and other Carlin-type deposits have made Nevada a leader among gold-producing states and have helped the United States become a major gold-exporting country.

As the two mine-tour buses filled with geologists snaked through narrow valleys leading up to the Carlin Mine, we could see the open pit where it all began, but on this trip we were headed further north to Barrick Gold Company's Goldstrike Claims.

Just north of Newmont's Carlin Mine the road enters a broad valley where mountains of heap leach pads define a new landscape. Barrick purchased the Goldstrike claims for $62 million in 1987, and did some deep drilling that led to the Betze Pit (named after geologists Keith Bettles and Larry Kornze who discovered it). After our two tour buses passed through the fenced-in Goldstrike administrative area, we eagerly disembarked for a sack lunch and briefing in a classroom. Then we went out to an overlook above the Betze Pit. The mine's 190-ton haul trucks were slowly moving loads of rock from the pit to the mill. On the far wall of the pit we could see an anticline—the underground geologic structure that had trapped the ore-bearing fluids and created the deposit. Barrick's plan was to extend the pit into new mineralized areas. On the horizon to the north, we could see the headframe of the new underground Miekle Mine, which was being connected by tunnel to another underground mine, the Rodeo, just beyond the Betze Pit area.

Our group headed down into the pit for a closer look at the geology, and parked by an excavator that totally dwarfed the two buses. We straggled toward the deepest workings, wanting to see the latest area being mined, of course, and all along the pit wall, we hammered at the soft, black, carbonaceous rocks. Alteration by percolating waters had destroyed the original composition and little solid rock remained. Sulfides sparkled in the dark matrix, and streaks of white carbonate highlighted the sooty specimens I collected. Later, as the buses lumbered up out of the pit, our guide pointed out some of the faults that had been the conduits for hydrothermal fluids carrying the minerals.

That evening we took over the Elko Super 8 Motel, our hiking boots thumping along the long corridors. We assembled at the community center next door for dinner and a presentation about the geology of the next day's mine visit. We saw slides of the stratigraphic column, discussed the names and characteristics of the important carbonate strata, and asked questions about the geology. For many of the participants this was a chance to get a look into another company's mine, to see if their rocks were any different, and to learn more about the regional geology.

At 7:30 a.m. we were already driving out of town toward another mining area north of Elko–Jerritt Canyon. As the road crested at Adobe Summit, we could see the Independence Mountains ahead, lightly dusted with snow. At the Independence Mill entrance we stopped to collect hard hats before driving on. Whole sections of mountains had been scooped out to create the Marlboro Canyon, North and West Generator, Murray and many other open pit mines. Three mines were in operation, while others waited for development. This tour was looking at the surface indicators that had led to the discovery of a deep deposit called the SSX. Samples of drill core were passed around.

At a bend in the road we piled out of the buses for a presentation. A test on a crumbly dike had indicated a weak fifteen parts per billion gold. Even with such a low gold content, a geologist had gotten a drilling program approved because of the proximity to another open pit mine, and other dikes trending in this direction. Drilling on the dike found gold at 570 feet, and a deep ore deposit 2,200 feet long and 500 to 800 feet wide was eventually drilled out and brought into production. The SSX deposit represents the new Nevada style of deposit—sulfide-rich and underground.

On the bus ride back to Reno, most of us spent the time discussing the deposits we'd seen. Old mine sites along the route were pointed out. Greg explained his test of "Least Amazement," which judges geological explanations based on their simplicity. He felt the Roberts Mountain overthrust was still an unproven theory: its complexity required too much "amazement."

NEVADA GEOLOGY

For those prospectors and rockhounds wishing a deeper understanding of Nevada geology—primarily recent, near-surface precious metal deposits–the Geological Society of Nevada (GSN) provides an entry into these mines. Dues are $25 a year, and the weekend field trips usually cost a few hundred dollars for registration, guidebook, meals and hotel, and transportation from Reno and return. GSN publications of past conferences and field trip guidebooks are also available and are reasonably priced and invaluable. Write to:

GEOLOGICAL SOCIETY OF NEVADA
P.O. Box 12021
Reno, Nevada 89510-2021
(775) 323-3500 / (775) 323-3599 Fax
E-mail: gsn@mines.unr.edu

NEVADA PUBLICATIONS
P.O. Box 15444
Las Vegas, Nevada 89114

A welcome bonus of the field trips is the attitude of the visited mines toward collecting samples. I've seen many chunks of glittering sulfide-rich rocks get carried away by visitors, and I usually fill several cotton sample bags, noting the mine name and rock types on the yellow tag on the outside of the bag, so I can study the deposit later, at home. The rock samples, together with the geologic reports, teach me about the deposit stratigraphy, favorable beds for mineralization, and types of alteration. Even though the disseminated gold isn't visible, I know that these rocks are examples of the new Carlin-type of gold deposit.

10. THE OLD AND THE NEW USGS

ALASKA USGS

I found Florence Weber, one of Alaska's first woman geologists, inside a white frame building with blue trim, on the University of Alaska Fairbanks' West Ridge. "I have fifty years of stuff around here," Weber admitted. "But I haven't forgotten most of the places I walked during my field work. I met a pack of wolves once. I only had a small gun, but I walked up to them and told them they were just a bunch of mangy dogs." She smiled at the memory. "They ran off. That day the helicopter didn't pick me up, and I ended up hiking twenty miles. I had to wade the Forty Mile River. But I waited for daylight before I did that. I could hear the wolves howling off in the distance."

USGS investigations in Alaska began during the 1890s when the Canadian Klondike and Nome, Alaska gold rushes were taking place. Geologist Alfred Brooks dropped his studies in Paris to accept work with the Survey in Alaska. In 1898 he joined gold rushers packing supplies over the new White Pass Wagon Road, and went down the Yukon with canoes to the White River. The party lined the canoes up the White River, wading continuously and fighting quicksand. They portaged five miles over the mountains into the Tanana Basin, where they examined the geology, completing their geological reconnaissance just as their supplies ran out.

In his summary of the gold resources, Brooks identified gold veins in Interior Alaska's Tanana schists as the source of gold. "We have seen that traces of gold have been found throughout the region examined by our party," he wrote. "I believe in spite of the adverse results which have been obtained so far, which are purely negative, that the White and Tanana River basins still offer a favorable field for the intelligent prospector."

These comments were made four years before the Fairbanks placers were discovered, and a hundred years before the rich Pogo Gold Deposit came to light.

"Alaska is eminently not the place for the haphazard or untrained prospector," Brooks concluded. "In the long run only those who have the intelligence, training, and patience to study the conditions of the occurrence of gold can hope to succeed in this district, where time is too valuable to allow of the happy-go-lucky methods which have been successful in other gold regions where the climatic conditions were more favorable."

Alfred Brooks went on to head the new Division of Alaskan Mineral Resources of the Survey, spending several months in the field every season, and publishing annual reports about mining in Alaska. During World War I he served in France, coordinating geologists who mapped terrain and found water sources for the troops. He died suddenly in 1924 after visiting several Survey field parties in Alaska. The Brooks Range, which separates the Arctic Slope from the interior, is named in his honor.

WESTERN REGIONAL CENTER

To see how the United States Geological Survey has changed, I recently traveled from Alaska to the Western Regional Center of the USGS in Menlo Park, California. Located near trendy Palo Alto (which has changed quite a bit since I grew up there), the landscaped grounds have signs directing visitors to numerous locations, including the map sales office in the Earth Science Information Center. In another building nearby is the scientific publications library, and upstairs are offices for the Water Resources Division. From her office, located in one of the many other buildings on the grounds, Pat Jorgenson, an outreach coordinator, explained the new USGS.

"Our original mission under John Wesley Powell was to map the surface and subsurface geology, and we have fulfilled that mission," Jorgenson told me. "The USGS continues to provide maps and geological information. However, now the focus is on water quality and availability, natural hazards, environmental contamination, energy and non-renewable resources."

"The USGS began one hundred and twenty years ago, in 1879," Jorgenson said. "The national center in downtown Washington, DC, was moved to Reston, Virginia, near the Dulles Airport, where the Eastern Regional Center is also located. The Central Regional Center in Denver serves the Great Plains and Rocky Mountains. Everything west of the Rockies is done here in Menlo Park.

"The USGS Western Regional Center was created after World War II," Jorgenson continued. "The government had army bases and land. The Menlo Park site was an army hospital, Dibblee Army Hospital, sixteen acres of land near Stanford and not far from the University of California at Berkeley."

The Western Regional Center of the USGS has the largest earth science library on the West Coast. Named for Vincent McKelvey who led the USGS search for uranium and was the ninth director of the Survey, the modern building contains USGS bulletins dating from 1884, professional papers, Canadian Geological Survey bulletins, maps, and an extensive collection of science journals. Large tables in alcoves are designed for map

The Western Regional Center of the USGS at Menlo Park, California has an Earth Sciences Information Center that sells maps, reports, and educational materials. The other USGS centers are located in Denver, Colorado and Reston, Virginia.

study. Except for the journals and reference materials, local libraries may borrow books through interlibrary loan. There are over 300,000 books, 35,000 maps, and 3,000 journals, plus special materials in the photo library, the California Information Center, and the Education Collection. The pleasant atmosphere, open stacks, and study areas make this library a great place to visit.

In a display case in the library building I saw paper models of spreading sea floor, crinoids, and volcanoes. I also noticed educational materials on the Antarctic ice sheet, sand dunes, the Chicxulub impact crater in the Yucatan, ocean trenches, making paper fossils, and other topics kids would enjoy. These and other K-12 education resources are on the education pages at http://www.usgs.gov/education/.

Nestled among native plants and flowers in front of the map sales building are three large granite rocks weighing several tons each. A sign identifies them as "granodiorite monoliths from near Sonora Pass in the Sierra Nevada that crystallized about eighty-six million years ago during the Cretaceous period." Black and gray squirrels chatter in the Monterey pines and oak trees.

"This center reflects its geographical position," Jorgenson explained. "We do water research that other agencies like the EPA rely on. We've got the oldest database on stream flow and water quality. Other areas of importance are earthquakes and volcanoes."

As I walked around I found other rocks: orbicular diorite, pegmatite, blue schist, and quartz from the California Mother Lode. Jorgenson led me to a quiet walkway to see breadcrust lava from the Mount St. Helens caldera explosion. A plaque beside it honored David Johnston, a geologist who was monitoring the volcano when it exploded May 18, 1980.

"It was Sunday morning and the other geologists had gone into town," Jorgenson said. "[Johnston] got on the radio and called headquarters: 'Vancouver, Vancouver, this is it!' They thought it would blow straight up, but it blew out the north side right toward him, killing him instantly."

JOHN "BEAVER" MERTIE

Volcanoes, water quality, and earthquake studies reflect the changing direction of the USGS. However, I still utilize the earlier USGS reconnaissance mapping and reporting on Alaska that was done by Survey men like John "Beaver" Mertie. Alfred Brooks hired him in 1911, and Mertie learned from the experienced geologist, Louis Prindle, how to pack horses and survive in the Alaskan wilderness. That first summer Mertie had to track down his wandering pack horses, bluff miners about his knowledge of gold placers, live with wet boots, and shoot a marauding black bear.

He worked in Alaska from 1911 until 1942, crossing the Brooks Range with dog teams one winter in order to do a reconnaissance of the North Slope in the spring and summer. The second week of the trip he wanted to wash so badly that one morning he went out -10°F weather and rubbed snow on his face and hands. Indians called him "Moose" because he often swam across rivers and lakes.

Funds were tight for the 1933 season, so he carried a seventy-pound pack containing tent, blanket, pots, compass, barometer, pick, gun, food and camera. He hiked 250 miles between mining camps, from Ruby to the Iditarod area. "It was a miserable trip," he wrote. "I was walking in mud almost up to my knees, the mosquitoes were unbelievable and I was black from head to foot with soot from the [forest] fires. The walking was so difficult that I couldn't make much more than ten miles a day."

His wife, Evelyn Mertie, later collected his notes and wrote up these experiences in *Thirty Summers and a Winter*, published by the Mineral Industry Laboratory at the University of Alaska. His primary work had been geologic mapping and examination of gold and platinum placers, but during the last thirteen years of his life he worked in southeast United States studying granitic rocks and minerals of economic value. Mertie worked for the USGS for seventy years and published forty-three reports about geology and mining in Alaska. His *Professional Paper 630*, "The Economic Geology of Platinum Metals," is a classic.

SECOND GENERATION GEOLOGISTS

Alaskan Florence Weber characterizes herself as part of the second generation of geologists after Alfred Brooks and John Mertie. She drove up the Alaska Highway just after it opened to civilians in 1949, with Florence Collins, another recent college graduate. "They called us the 'gold dust twins.' We worked on the Navy drilling project mapping drill cores from the Naval Petroleum Reserve Number 4 on the North Slope. We came to Alaska because we wanted to go out in the field, but the Navy wouldn't allow us to go to the North Slope. That made us mad. We worked out of a Fairbanks office instead.

"Once the Navy operation was closed down, we were transferred to Washington, DC. They dragged us out of Alaska kicking and screaming. We soon got a grant to do field work on the Minchumina Sand Dunes near Mount McKinley. We both flew, and we decided to get a float plane and fly it north to use on the job. It was just a Super Cub with floats.

"We learned how to fly with floats in Syracuse, New York, and then started our trip from Chesapeake Bay. In 1956, it was a big adventure to go from Washington, DC to Alaska that way. We went by Cleveland, Chicago, Milwaukee, and crossed into Canada near the Lake of the Woods area. We flew north along Lake Winnipeg, Flin Flon, Lac La Ronge, Fort McMurray, and north to Yellowknife across the Great Slave Lake. It was unusual for two gals to be flying a float plane in those days. We ended up at the mouth of the MacKenzie River and landed on the Arctic Ocean. After looking at some pingos created by permafrost, we went on to Old Crow, Fort Yukon, and finally Fairbanks. Florence Collins eventually published an open-file report on the sand dunes and surficial geology of the Minchumina area.

"After I returned to Washington, they were just starting the Menlo Park Regional office and wanted to transfer me there. I was fortunate that Troy Pewe offered me a job as his assistant in 1957 in Fairbanks, and I've been in Alaska ever since. I'm a generalist. Paleontology was my speciality at the University of Chicago, then I got into surficial geology, then stratigraphy and structure. I can change hats and map surficial as efficiently as hard rock.

"The Tintina Fault [in Alaska and the Yukon] has always been one of my enthusiasms," Weber admitted. "The standard displacement along it has been thought to be 450 kilometers [279.62 miles]. That's what the Canadians talk about, but they're looking at the body of the elephant. In

Alaska there seems to be a lot more movement–possibly 700 to 900 kilometers [435 to 559 miles]. We're the head of the elephant."

When I later visited the USGS office in Menlo Park, Jorgenson told me that the Alaska Branch of the USGS is no longer as large as it once was. "Some geologists are completing geologic maps, and the Anchorage office has a team doing field work on more than forty active volcanoes," she noted. "The USGS is a partner in the Alaska Earthquake Information Center [at the University of Alaska Fairbanks' Geophysical Institute] in Fairbanks, which performs round-the-clock earthquake recording and analysis and disseminates information. [Department of] Water Resources people are active statewide. We are participating in a digital mapping program for the entire state."

THE NEW USGS

The USGS continues to publish informative bulletins and professional papers. A recent series on gold deposits covers the geology and resources of gold (USGS *Bulletin 1857A*), gold terranes in the United States (USGS *Bulletin 1857B*) and descriptions of major gold deposits types (USGS *Bulletin 1857 C,D,E,F,G*). However, the Survey has changed to meet the concerns of today's world. Their studies reflect an increasing awareness of the environment. Mapping is shifting to digitally-based data, aerial photo maps are adding a new dimension, and web sites allow public access into many aspects of the USGS. Tectonic interpretations are now part of geological maps. The threat of volcanic eruptions is closely monitored. Faults are mapped and studied as living parts of our geology. Water quality and stream flow have taken on new significance.

"Private companies are producing many maps now," Jorgenson explained. "We still do the research here and provide the data, but everything is now done by computer, and the Denver Center handles map revisions. A new product we're doing is photo quadrangles from aerial surveys, at the same scale as 7.5-minute topographic maps. These are useful for land use planning. They show hills, roads, and vegetation types."

Inside the Earth Science Information Center (ESIC), I browsed through shelves of topographic maps of the western states. They reminded me of my first use of USGS publications. When I was fifteen, my two older brothers and I bought topographic maps to use while hiking California's Sierra Nevada Mountains. After I moved to Alaska and began prospecting, I turned to USGS bulletins for information on Fairbanks and other gold-producing areas.

Carousels inside the entrance to ESIC held USGS Fact Sheets about volcanic eruptions, earthquakes, placer mining in Alaska, new earth science data bases, and recent USGS highlights about natural resources, the environment, and hazards. "Reports are still being produced," Jorgenson said. "But rather than bulletins, which take years to do, we put out fact sheets. These are a page to six pages in length rather than a full report and take only three to four months to produce. They are still peer reviewed for accuracy, but these are more timely."

I approached the ESIC counter, asked for information on a specific topic, and a data base in the computer was pulled up. The USGS has many web sites, saving people the trouble of coming to the regional center. Their home page and geologic information page is http://www.usgs.gov/.

When I got back to Fairbanks, I immediately looked up the USGS list of bulletins and professional papers and ordered 1940s studies on the Black Hills (USGS *Professional Paper 247*) and New England pegmatites (USGS *Professional Paper 255*), a report on jasperoid (USGS *Professional Paper 710*), and a new bulletin about the gold placers of the Forty Mile area of Alaska (USGS *Bulletin 2125*). These can be located on their web site at: http://greenwood.cr.usgs.gov/formal/reports.html. The Earth Science Information Center can be found at: http://ask.usgs.gov/.

LASTING EFFECTS OF THE USGS

I recalled my visit with Florence Weber in her cluttered Fairbanks' office, where she sat before her computer and reminisced about her field work in Alaska. "I've tramped over a lot of the Yukon-Tanana uplands. I went over the ground where they're making some big discoveries. Our original mapping must have guided them. I remember walking down those ridges where the Pogo gold discovery took place. I'm glad to see it's being developed.

"There were the old-time geologists who laid out the outline. Then we came along and did the quadrangle mapping. We went out and found out what the rocks were, and did stream sediment sampling. Now this is a third generation. They're specialists working on the details."

As I was leaving Florence Weber's office I noticed a faded geologic map in the entryway showing the Fairbanks Quadrangle. It represented the view out her window to the snow-covered Alaska Range—showing the

Fairbanks gold-bearing hills, the vast gravels and floodplain of the Tanana Valley, and the distant slopes of Mounts Hess, Hayes, and Deborah. Weber's name was listed as one of the creators of the map.

A year later, when I was driving past where her old office had been, I realized the building was gone. The smooth carpet of snow between the spruce trees showed no sign of the place where so much had been accomplished.

The field geologists such as J. Beaver Mertie, Alfred Brooks, and Florence Weber, who covered the country on foot, rafts and boats, horses, and helicopters, began an enormous task of geological mapping. Their gold rush reports and maps have a special place on my bookshelf. These classic studies of famous districts continue to provide insights and information for prospectors to use, in addition to the more recent maps, bulletins, fact sheets, and reports about the geology of an area. Behind the maps and reports are dedicated people who convey a sense of the romance, history, and human understanding of geology.

HOW TO REACH THE USGS

USGS Information Number (888) ASK-USGS (888-275-8747)
http://www.usgs.gov/

EASTERN REGION AND HEADQUARTERS
USGS National Center
12201 Sunrise Valley Drive
Reston, VA 20192
(703) 648-4000

CENTRAL REGION
U.S. Geological Survey
Box 25046 Denver
Federal Center
Denver, CO 80225
(303) 236-5900

WESTERN REGION
U.S. Geological Survey
345 Middlefield Road
Menlo Park, CA 94025
(650) 853-8300

To locate major U.S. Geological Surveys across the country, visit
http://www.usgs.gov/major_sites.html

For more information regarding the USGS Visitors' Center in Reston,
Virginia, call (703) 648-4748 or (703) 64-VISIT, or visit their website at
http://mapping.usgs.gov/mac/visitors/html/intro.html. This center
provides an opportunity for visitors to experience and learn about natural
science through tours and exhibits. This center is for community groups,
professional societies, school classes, and families. Guided tours are available
through appointment only.

11. RESEARCHING A PROSPECT

Looking at a status plat of state mining claim locations north of Fairbanks, Alaska, I was struck by a line of placer claims extending up the valley. Obviously the placer miner had drilled the valley and found gold. Why else would he hold on to those claims and pay the yearly fees? Looking over the results of a state geological sampling program, I found increased arsenic levels near the head of the creek. Then I remembered seeing thick quartz veins during a reconnaissance hike around those hills. Quite possibly, I had researched a lode deposit into life.

During the off-season, tracking down references in libraries, state mining offices and surveys, and federal agency publications generates new ideas for finding mineral deposits. A one-sentence note in an old USGS bulletin may hold the clue to a new mine. I once mentioned a brief description about iron pyrites and gold around Wickersham Creek in the Alaska Range to a mining company geologist.

"We spent a season on that," he admitted. "Brought in a drill rig and put in a few holes. We probably should have done more."

We had both seen the reference in an old report. On such hints a whole season's program may be built.

USING THE LIBRARY

Another kind of research takes place when I run into a strange suite of minerals in an assay report or recognize a new deposit type. I recently began working in an area with turbidites, undersea landslides that formed in shallow basins, and I discovered they were often associated with gold-quartz veins in Canada. Looking through Canadian Geological

Survey *Bulletin 540*, "Geological Classification of Canadian Gold Deposits" (2000), I found frequent mention of this deposit type. Similarly, when I was investigating a gold placer and I came across a lamprophyre, an igneous rock from deep crustal sources, I found they were often associated with gold, so I studied up on them.

Unfortunately, my local public library in Fairbanks has only a small section on geology, and an even smaller selection of books on mineral deposits. Volcanoes, dinosaurs, and fossils are the most popular subjects. Thin books on gold panning take up the most shelf space. In the western americana section there are a few books about the gold rushes to California, Nevada, Montana, and South Dakota. Although many public libraries have interlibrary loans and Internet access, doing geological and mining research there is still very limited.

WHERE TO BUY CLASSIC GEOLOGY BOOKS

If I want books on economic geology, I visit the University of Alaska Fairbanks' Rasmuson Library. Here, I can check out the classics: Waldemar Lindgren's *Mineral Deposits*, William H. Emmons' *Gold Deposits of the World*, Alan M. Bateman's *Economic Ore Deposits*, and Charles F. Park and Roy A. MacDiarmid's *Ore Deposits*.

These and other texts came out of the early- and mid-1900s when geologists were building mineral deposit classification schemes, and now book dealers sell these at bargain prices. I order used and rare geology books from:

SIERRA BOOKS
P.O. Box 2504
Martinez, California 94553
(925) 228-1849

MT. EDEN BOOKS AND BINDERY
P.O. Box 1014
Cedar Ridge, California 95924
(916) 274-BOOK

or search the internet, if I have a book title, at www.bibliofind.com.

New books on the subject are expensive–Mead L. Jensen and Alan M. Bateman's *Economic Ore Deposits* (John Wiley and Sons) costs over $70–but they incorporate new deposit models and ideas from plate tectonics. Robert Boyle's "The Geochemistry of Gold and its Deposits" (Geological Survey of Canada *Bulletin 280*) synthesizes the important literature on gold. My university professor had his ore deposit students buy the sturdy paperback *Ore Geology and Industrial Minerals* by Anthony M. Evans (Blackwell Science. Third Edition, 1996). These general background references are invaluable for reading up on deposit settings and characteristics.

PRELIMINARY INVESTIGATIONS

When prospecting a particular area, investigating what has been done there is an essential first step. This would include looking over maps, aerial photos, reports, studies, status plats, and old mining claim information.

"If you are way out in a deserted wilderness, and you come across a bunch of named creeks on a USGS topographic map," Erik Hansen, a geologist-researcher, confided, "the names indicate that in the early days there was enough encouragement for old prospectors to dig prospect holes." He pulled out a topographic map of the Big Delta quadrangle southeast of Fairbanks, Alaska. "Pogo [a world-class gold discovery] is right over here." His finger pointed to a remote area. "You don't see a lot of creek names–Sonora Creek, California Creek. Just a few small creeks got a name. Recent prospecting around the Pogo area found that, yes indeed, there is gold in Sonora Creek."

He bent over and surveyed the area. "Uncle Sam Creek. That sounds good. You have Monte Cristo Creek. Right across from that is Gold Run Creek. There has to be a reason for this and there is." He paused, smiling through his neatly trimmed beard. "After the Pogo area was claimed, people began to concentrate over here. Cominco found gold, but before that old-timers found gold. It will be thoroughly evaluated over the next few years."

Erik Hansen, geologist and consultant, researches land status for mining companies. Before prospecting, it is important to check land status and determine if the land is open to claim staking.

We had met for an interview over a table in the Division of Mining room of the State of Alaska Department of Resources Building in Fairbanks. Prospectors, miners, and mining companies come here to search out mining claim information, pay for new claims, and get claim forms. Hansen, once an employee with his own desk in this room, now spends time searching out land status and claim ownership as part of his consulting business.

"The USGS Dictionary of Alaska Place Names is another interesting source of information on mining activity," he told me. "Old timers out in the hills named little creeks where they were prospecting. The Dictionary can tell you when it was named." He thought of another clue. "USGS shaded relief maps have more old trails and cabins than the more modern topographic ones, and old cabins are great because they are either trappers cabins or prospectors cabins."

A trapper would place his cabin in the river valley where frozen ground in winter allowed easier travel to trapping sites in the valley, while a prospector would build his cabin along a placer creek, usually a side creek where bedrock was closer to the surface. Prospect holes and workings near these cabins yield good samples of bedrock and creek gravels.

RESEARCHING LAND STATUS

Hansen had done his share of field geology staking claims in remote areas, but now he researches land status for mining companies. Always ready with a mischievous smile, he reminded me of a lean cross-country skier. Taking time out from his work, he explained the procedures he follows when researching land status.

"First we need to define the area. I need a list of townships that defines where this is. Then I would determine whether it is state, federal or private land–and whether the federal or state land is open to mining location. At the BLM, the status plat will show status as PLOs [Public Land Orders]. Here at the State Division of Mining I look for MCOs [Mineral Closing Orders] and the Area Plans–for example, the Tanana Valley Basin Plan that will tell you if it's open or closed."

Flat gray drawers with microfiche of each township in Alaska stand nearby. I often pull up a township I'm interested in to check who owns the land and to see where state mining claims are located. That's how I found the line of placer claims in the creek valley that suggested a potential gold source near the headwaters.

"For those state areas that are open," Hansen continued, "I will obtain a status plat. Unfortunately, the status plat is outdated [because of the time lag between filing a claim and its incorporation into the plat]. It also doesn't show [State of Alaska] prospecting sites. My next step at the state level would be to print a land abstract of those townships. This will be a little more current." Two computers stood ready to print out this information. "For those active claims, I will check to see if the claim rent was paid and if the annual labor was timely reported. For the prospecting sites, I will check to see if the sites have expired or been extended. Then I would go to the recorder's office to see the most recent recorded mining activity in that area."

He snapped open his leather briefcase and pulled out a map he had just done. Interlocking rectangles with numbers keyed to mining companies, and yellow and red blocks indicated important claim holdings. This particular piece of Interior Alaska was an active area with a lot of interest.

"I indicate recording district boundaries on my map so the client will know where to record location notices. If it's a new out-of-state client, I will make sure he is aware that before he can stake a valid mineral location in Alaska, he must be registered to do business in Alaska. I give them a copy of the rules and regulations on mining claims." He flapped a thick booklet of typed pages in the air. "It's actually crammed full of good information, but it can put you to sleep."

Since open areas of western states are federal lands, these have federal mining claims. "Over at the Bureau of Land Management, we don't find federal mining claims plotted on status plats," he noted. "So I use the BLM computer terminal to print out abstracts of active federal claims in the area of interest. Then, I either ask for their case files or use the old State Kardex system." Combining his research from the Alaska Division of Mining and BLM, he has information on all the active state and federal claims in a particular area of Alaska. Based on a map he draws up, a mining company knows exactly where the open lands are for investigating and staking.

To get a picture of where information on federal mining claims is maintained, I drove to the pink granite BLM building. Susan Rangel, Land Law Examiner-Public Contact Representative, greeted me from her computer station in the Public Room.

"If they want to know if it's open for staking," she explained, "we'd pull a master title plat to find land ownership and mineral entry status. We'd need the township and range and meridian and section. We'd put that land description in the computer and see what mining claims would be in there. Mining claims would show up from 1979, when the Federal Land Planning Management Act required miners to file with us.

"The computer issues a serial number for claims. That's the case file. Anything you file goes into that case file. That's public information: when the location notice was first filed, when the discovery was made, annual assessments–any paperwork that's filed."

In the past I've requested case files on some creeks and found crudely drawn maps, deeds, and the annual labor reports. Filing cabinets at the State Division of Mining contain similar paperwork and are a great source of information about what mining work was done each year. Checking on some state claims, I learned how much drilling had been done from the miner's annual labor report. Since the claims were dropped the year after drilling was done, I decided the drilling results must have been negative.

Once open areas for prospecting are identified, favorable geology becomes a limiting criteria. USGS reports provide some information, and states with a lot of mineral activity, such as Arizona, Nevada, and Alaska, usually produce their own geological publications, maps, and studies of mineralized districts.

CHECKLIST FOR RESEARCHING AN AREA

1) Get a **topographic map** of the area you are investigating. **1:63,360 scale** is best.

2) Locate the **townships** and **ranges** that define the specific area being considered for prospecting. Find out the **meridian** so your legal description on the claim form is complete. The recorder's office will know the meridian.

3) Obtain a **land status plat**. State land offices and the BLM offices will have this information. The land status tells you whether the prospecting area is state, federal, or private land. Many areas are closed to staking mining claims. This is the **most important step** in order to avoid wasting time on closed areas.

4) Using the townships and ranges for your area of interest, obtain a **print-out of existing mining claims**. Mark their location on your map. If it is state land, you will need a list of **state claims**. If federal land (BLM property), you will need a list of **federal claims**. Since some of Alaska's land has changed from federal to state ownership, some people still have federal claims on state lands, so a list of both kinds of claims is necessary to avoid claim staking conflicts.

5) Check records in land offices to see if **claims are active**–e.g., paid-up annual fees and doing timely Annual Labor Reports.

6) Check state recorder's office for **recent mining claims activity** in the intended prospecting area. New claims will not always be listed in state and federal land offices until several months after being recorded.

7) Get **blank claim forms**. Read state and federal rules and regulations for staking a claim on the ground, and for keeping it active. Once learned, yearly filing requirements and fees are easy to follow.

8) If staking in Alaska, be **registered to do business.**

9) Work with other surface owners such as county or borough land managers to ensure proper access and other permitting when necessary. Reclamation is an important issue. Keep it in mind at all times.

10) Prospect the area.

11) If an economic deposit is found, stake the claim(s), and plan on more extensive analysis of the ground by **surface sampling** (soils, rocks, stream concentrates, and sediments), trenching, and drilling. Any surface disturbance beyond soil, stream panning and rock sampling will require permits.

OTHER SOURCES OF INFORMATION

The Arizona Geological Survey has a list of bulletins, circulars, maps, and open-file reports. For more information, write to:

THE ARIZONA GEOLOGICAL SURVEY
416 West Congress Street, Suite 100
Tucson, Arizona 85701
(520) 770-3500 / (520) 770-3505
http://www.azgs.state.az.us/

Bulletin 137, "Arizona Lode Gold Mines and Gold Mining" and *Bulletin 168*, "Gold Placers and Placering of Arizona" cover many important regions.

Gem Guides Book Co. also publishes titles such as *Placer Gold Deposits of Arizona* and for the states of Nevada, New Mexico, and Sierra Nevada.

GEM GUIDES BOOK CO.
315 Cloverleaf Drive, Suite F
Baldwin Park, CA 91706
www.gemguidesbooks.com

Similarly, the Arizona Geological Society puts out the *Arizona Geological Society Digest Volumes*, proceedings from symposia held at the University of Arizona in Tucson. For more information, write to:

THE ARIZONA GEOLOGICAL SOCIETY
Publications
P.O. Box 40952
Tucson, Arizona 85717
(520) 663-5295
http://www.arizonageologicalsoc.org/

These volumes like, "Frontiers in Geology and Ore Deposits of Arizona and the Southwest," contain many useful papers and field trip guides.

Whenever I visit Reno, Nevada, I spend time at the University of Nevada where the Nevada Bureau of Mines and Geology has a publications and map room in the Scrugham Engineering-Mines Building. For more information, write to:

NEVADA BUREAU OF MINES AND GEOLOGY
Mail Stop 178
University of Nevada
Reno, Nevada 89557-0088
(775) 784-6691 ext. 133 / (775) 784-1709 Fax
http://www.nbmg.unr.edu/
E-mail: nbmginfo@unr.edu

They publish bulletins on mining areas and important minerals, placer mining in Nevada, histories of famous districts, and geologic maps.

Located downstairs in the Laxhalt Mineral Engineering Building on campus, the Geological Society of Nevada has symposia proceedings from several major meetings on deposits of the Great Basin and American Cordillera, as well as field trip guides to important mining centers. For more information, write to:

GEOLOGICAL SOCIETY OF NEVADA
P.O. Box 12921
Reno, Nevada 89510-2021
(775) 323-3500 / (775) 323-3599 Fax
gsn@mines.unr.edu

Nevada Ghost Towns and Mining Camps by Stanley Paher contains information and black and white photos of Nevada's old camps. This and other mining and geology books are available from:

NEVADA MINERAL AND BOOK COMPANY
P.O. Box 44230
Las Vegas, Nevada 89116
(702) 453-5718

A reprint of Francis Church Lincoln's 1923 bibliography, *Mining Districts and Mineral Resources of Nevada*, is also published by the Nevada Mineral and Book Company.

VALUE OF EXAMINING USGS BULLETINS

Back in the 1890s, USGS expeditions to Alaska began reconnaissance mapping and reporting on the new gold discoveries. Geologist Alfred Brooks visited Nome a year after its gold rush started, and after viewing the new beach placer, he looked inland and saw the potential for other buried gold-bearing beaches.

"Strangely enough," he later wrote, "though the facts and deductions were presented in a widely distributed publication which was in the hands of thousands of the gold seekers of 1900, so far as I know not a single one paid any heed to the search for ancient beaches." Another USGS bulletin the following year went into more detail, but still no one followed up his suggestion. Four years later the Second, Third, and Fourth Beaches were discovered. The value of carefully reading USGS bulletins was finally recognized.

To help people locate information on a particular area in Alaska, Edward H. Cobb and Reuben Kachadoorian compiled references to reports from federal and state agencies through 1959 in USGS *Bulletin 1139* (1961). Within each quadrangle the index lists creek names, mining companies, and the names of property owners followed by the references where the entry was found. Cobb also compiled "Placer Deposits of Alaska" (USGS *Bulletin 1374*) and with Henry C. Berg he did "Metalliferous Lode Deposits of Alaska" (USGS *Bulletin 1246*). These last two references are available from Alaska Prospectors Supply, 504 College Road, Fairbanks, Alaska 99701. Virginia Doyle Heiner's *Alaska Mining History: A Source Document* (Alaska Division of Parks, 1977) contains bibliographic references to many Alaskan deposits.

Published in 1999, "The Guide to Alaska Geologic and Mineral Information" (DGGD *Information Circular 44*) has addresses and web sites for all the important sources of geological information. For more information, write to:

THE DIVISION OF GEOLOGICAL AND GEOPHYSICAL SURVEYS, DGGS
794 University Avenue, Suite 200
Fairbanks, Alaska 99709-3645
(907) 451-5000 / (907) 451-5050 Fax
http://www.dggs.dnr.state.ak.us/

The Alaska Division of Geological and Geophysical Surveys, United States Geological Survey, Bureau of Land Management, and seven research libraries in Alaska produced the publication for geologists and mining companies.

"It was a very cooperative effort," Ellen Daley, editor of the publication, told me. "The librarians did a lot of research. It couldn't have been done without a large group of people behind it. The decision was made fairly early on to focus specifically on mining and minerals, and we pretty much stuck to things interesting to the mineral industry—all the way from the prospector to Teck Corporation."

Daley's office area was dominated by a large computer, scanner, and work tables. She flipped through the pages of the publication. "We wanted to put the Guide on the web as an updateable site. It's accessible on the DGGS website [go to http://www.dggs.dnr.state.ak.us]. The librarians are working on updates, and I'll go in there and fix them."

She is now scanning DGGS publications in order to place them online for people to download. The Division has a catalog of its publications, and a web site with a list of its publications on Alaska arranged by quadrangle.

"All the information in the world could be buried in a vault and it would be worthless unless you had access to it," Daley feels. She pointed to two stacks of DGGS publications ready for scanning. "They will be totally free. When they are accessed on the Internet, a person can either read them, down load them, or print them, including maps."

During this transition period from hard copy to the web, *The Guide to Alaska Geologic and Mineral Information* provides an overview of information sources to follow up, including sources of unpublished information, a directory of agencies, basic references for geology and mining in Alaska, recreational mining information, land status, and mining claims. Major publications by the USGS, U.S. Bureau of Mines, and Alaska DGGS also are listed, along with the agencies' web site addresses. Although this free publication focuses on Alaska, it describes much of the data available for researching mineral deposits in the United States.

Other sources of information include the USGS National Geochemical Database containing the Department of Energy's National Uranium Resource Evaluation Hydrogeochemical and Stream Sediment Reconnaissance (NURE HSSR) Program of stream-, lake-, and pond-sediment samples. The Earth Resources Observation System Data Center (EROS) has satellite data, and raw data from the NURE program (go to http://edcwww.cr.usgs.gov). Some libraries have GEOREF, a database of North American geologic literature from 1785 to the present. Scientific journals such as *Economic Geology, Engineering and Mining Journal,* and *American Mineralogist* are good sources for articles about new discoveries. I also consult the bibliographies to articles in these periodicals for additional sources to consult.

SUMMARY

As Erik Hansen emphasized, any search begins with a particular place. Deciding on a worthwhile target is a combination of intuition, fieldwork, and good research. In spite of the staggering amount of information available, effort and persistence can uncover useful material.

Mining companies are conservative and they generally begin with known mining areas. The independent prospector can step out more easily, selecting targets based on minor clues, trends, or accessibility. Taking time to do the homework can make the search more of a solution to a question than a stab in the dark.

12. The Lure of Lost Mines

Strange birds calling from the cactus and mesquite hillside behind our tent woke me as dawn was warming the sandstone cliffs. I walked in the shadows along a stony creek, enjoying the solitude and morning coolness. As I crouched beside a pool, I heard someone cough behind me. A gaunt man in loose jeans and a ragged flannel shirt had silently appeared without warning.

"You looking for the mine?" he shyly asked. "Most people camping here have heard about the Lost Dutchman Mine."

I was in the Superstition Mountains east of Phoenix, but this was the first time I had connected the legend with the locale. I quickly assessed my visitor: no gun, a beet-red burn on his forehead, and the look of a vagrant.

"My wife and I are just enjoying driving the Apache Trail," I told him. "My parents brought me here when I was a kid, and I wanted to see if anything had changed."

Legends of lost mines, like any treasure, attract a curious crowd of adventurers, crackpots, and believers. I've heard my share of stories involving a lone prospector, a remote discovery, sudden wealth, losing the way back to the mine, or death. Somehow, a rough map or description giving sketchy clues to the location survives, or the visible treasure the prospector displayed gives credence to the claim. Few lost mines are ever found, but this doesn't mean some aren't out there.

VALIDITY OF LOST MINE ACCOUNTS

Andy Harman, a Canadian mine-maker, related several stories he'd heard from reliable sources. He comes from a practical background where mines are judged against cold facts.

"When I was packing a string of horses into my guiding camp along the Canol Pipeline to Norman Wells, Yukon Territory, I went through one of those high passes that had the right environment for gold. I knew a guy who found placer gold there and Indians reportedly brought out gold from the area. I keep the location in the back of my mind for future prospecting," he confided.

"Another story I heard was about some guys who used to go into the Cariboo [country] in British Columbia in the spring and come out with heavy packs in the fall. I think they found some new creeks. You have to report all gold to the government, and I've seen those records."

For Harman, the criteria for chasing a lost gold mine include tangible proof of gold, a reliable discoverer, and a proven environment. It's human nature to want sudden success–a full-blown mine or a sure thing–but it pays to be realistic when assessing a story, because lost mines have a habit of being remote and hard to get to.

When I lived along the Yukon River, Nulato Indians told me stories of prospectors going off into the Kaiyuh Hills and finding gold. On the way out their canoe capsized and their outfit was lost. I heard another story from an Eskimo who recounted how his grandfather noticed shiny gold on a ridge east of the coastal village of Unalakleet while out herding reindeer, but he couldn't find the place again. Another Eskimo friend described finding a nugget on Alaska's North Slope.

I never bothered to follow up these stories. The canoe story would require superhuman efforts just to reach a range of low hills. I've been guided through the Kaiyuh Swamp by experienced Indians, and I didn't enjoy the hordes of mosquitoes and convoluted sloughs. The Unalakleet reindeer herder was out on sedimentary rocks where pyrites and coal were more likely. And the North Slope is composed of thick sequences of gravels from the Brooks Range. Nuggets would be widely dispersed from distant sources.

MOSES CRUIKSHANK

When the story comes from an experienced prospector, people pay attention. Moses Cruikshank, an Athabascan Indian from the Fort Yukon area, learned from old-timers how to sink shafts in Alaska's permafrost ground with a steam boiler. In his oral history, *This Life I've Been Living*

(University of Alaska Press, 1986), he described his years trapping and prospecting north of the Yukon River. He worked for the dredges around Fairbanks, and partnered with some of the best prospectors.

One fall while running winter supplies up the Hodzana River to his trapline cabin by boat, he noticed that the river sediment looked strange and he stopped to pan it.

"Of all the pans I've ever taken" he wrote, "I believe that is the richest pan that I ever took. After I pan it down, oh, the black sand on it was that thick, you know. I panned it down on a big standard size gold pan, and the rim of the gold pan for about ten, twelve inches, there it's just yellow like corn meal. You know, flour gold" (Cruikshank, p. 87)

I asked him in person about the story, and he launched right into the account again. He had to walk with a cane, and it was shortly before he died, but he brightened up at the memory.

"It was late in the fall, and I decided to wait until next spring to stake it, the spring of 1942. I didn't want someone else to find it before I had a chance to look it over," he explained. His huge hand slapped the table in exasperation. "That Christmas I found out about the war. I got drafted. When I got back, ten feet of muck covered everything!"

He trapped and prospected the area for many years, but he never relocated the rich placer ground. I checked a land status map and the Hodzana River is now within the Yukon Flats National Wildlife Refuge where mining is prohibited.

FELIX PEDRO

Another experienced prospector, Felix Pedro, found paying gravels on a tributary of the Tanana River a few years before he discovered the Fairbanks placers. He felt confident that he could find his way back to the strike. He dug his shovel into a prominent bald knob and hung his jacket over it before going downriver to the nearest trading post, but he never found the spot again, even though he tried several times. He always maintained it was as rich as Fairbanks, but no one ever located it.

PAUL SOLKA

Perhaps this was the same lost mine Paul Solka wrote about in his account, "Lost Gold Mine of the Upper Tanana" (*Fairbanks Daily News-Miner*, 1994). Solka's father heard about two men who prospected from the Yukon to the Tanana River. In 1900 they found a rich placer within sight of the Tanana River. One partner died and the other went out to Tacoma, Washington, and later drew a crude map for a saloon keeper before he died of pneumonia.

Solka's father mined in the Tenderfoot District seventy-five miles southeast of Fairbanks near the Tanana River, and his wife and son compiled all the accounts of the search for the lost mine. According to one story, a resident at the mouth of Shaw Creek, "Crazy Tony" (Solka, p. 9) Kozloski, found a rich vein of quartz in the hills, but he was institutionalized and never developed his find. He did, however, leave a sample of rusty quartz with "grape-like clusters of gold from BB shot to buckshot" (Solka, p. 21-36) at the local store and roadhouse.

Chief David Charley of the Goodpaster River Tribe of Athabascan Indians drew a map of where he'd seen miners working a claim around Eagle Creek off Eagle Dome. Solka's father and some others packed into the area and dug several shafts, but the overburden was forty feet deep and they couldn't find the gold-bearing channel. Solka figured Kozloski's quartz vein was located off Eagle Dome, and that it had eroded and formed the placer the earlier prospectors had worked. This area is only about twenty-five miles from the recently discovered Pogo Lode Gold Deposit, a sequence of thick quartz veins that lie almost flat. Needless to say, the whole area is being intensely prospected, and the lost mines of Felix Pedro and "Crazy Tony" (Solka, p. 9) may still be found.

WALT WILHELM

Walt Wilhelm's gripping story of his father's prospecting in the western states in *Last Rig to Battle Mountain* (William Morrow and Co., 1970) mentions several lost mines. Sherman Wilhelm may have found early indications of the Nevada Carlin Trend when he investigated the Tuscarora Mountains and found gold "a few miles west of Maggie Creek," (Wilhelm, p. 230) just north of the town of Carlin.

Wilhelm took his family across the Sierra Nevada Mountains using the Sonora Pass in order to search for the Lost Mexican Mine on Dead Man's Creek. This was only a pack trail in the 1860s when a Mexican lost a pack animal on a steep stretch of trail. While retrieving the contents of the pack, he picked up quartz with streaks of gold. Many people in Bridgeport, California saw the specimens and this may have led to his murder. Wilhelm investigated the rocky gorges around Sonora Pass, but eventually gave up his search and moved on.

In Austin, Nevada, the Wilhelms heard about the Lost Breyfogle Mine. A wagon maker with a small shop in Austin in the early 1860s had found a flat-lying vein of reddish quartz containing nearly solid gold. Wilhelm quoted locals as saying, "The ore was so full of gold, it glistened in the sunlight like a polished gold frying pan" (Wilhelm, p. 234). People in Austin saw specimens of the rock, but all they knew was that it was located less than two days ride from town. Breyfogle lost the location and he went crazy searching for it.

I checked my mining and prospecting texts for how lost mines should be treated, but the topic wasn't seriously addressed. C. Godfrey Gunther rediscovered the lost copper mines of Cyprus through his close analysis of Biblical texts. In Gunther's *The Examination of Prospects* (McGraw-Hill, 1932), he evaluated known deposits, and he felt that in any mining examination, the "owners generally suffer from extreme optimism" (Gunther, p. 5). This seems true of lost mine stories as well. The placer is always located right on bedrock and is easily worked. The quartz vein's richness precludes any thought that it might be superficial and not extend to great depth. Like many a fish story, an objective view may be lacking.

USEFULNESS OF LOST MINE STORIES

William Peters' comprehensive *Exploration and Mining Geology* (John Wiley and Sons, 1978) credits prospectors with having a wealth of valuable information "even though it will be from an optimistic viewpoint." All geologic information has an element of subjectivity, he concedes. Peters feels that many successful exploration programs should begin with a search of the literature. While he doesn't mention lost mine stories, even these are useful if their source is credible, as in Moses Cruikshank's account. Peters' believes locale is also important: "Most mineral deposits are discovered in districts where someone has mined before, where an early geologist noted something of possible importance, or where some long-forgotten prospector filed a mineral claim" (Peters, p.282).

In his bibliography, *Lost Mines and Buried Treasures of the West* (University of California Press, Berkeley, California. 1977), Thomas Probert lists published accounts of lost mines from Kansas to Texas to California. In his analysis, these stories are usually told in the first person, gaining credibility because of known events, places, and people. Natural disasters usually lead to the disappearance of a mine or treasure. The stories are passed on through oral or written accounts. He feels that some of the lost mines and treasures may have been found, but the discoverer kept it a secret to avoid attention.

Emerald deposits in Ecuador were rediscovered by Stewart Connelly, who searched the colonial records in Madrid, Spain and in Quito, Ecuador. He noted that Pizarro had received seven large emeralds from Atahualpa, the last Inca emperor in this area. After a nine-month trip, Connelly successfully returned with emeralds, one weighing twenty carats. Unfortunately, on his second expedition in 1924, he disappeared. Connelly investigated a known emerald locality and found the gems, but he apparently died in the effort.

LOST DUTCHMAN MINE

Erle Stanley Gardner wrote *Hunting Lost Mines by Helicopter* (William Morrow and Co., 1965), an enthusiastic story about searching the Superstition Mountains in Arizona for the Lost Dutchman Mine, and near Yuma, Arizona for the Lost Nummel Mine. John Nummel, an experienced miner in the Red Cloud and La Fortuna Mines around Yuma, had been resting in the shade of a verde tree off the Yuma Wash when he scratched a dirty yellow rock strata of almost pure gold. (The geology sounds questionable, since visible alteration effects should extend into the surrounding rocks.) Nummel didn't take a sample with him, and was unable to find the spot, but the story may have furnished him with a grubstake from gullible listeners.

Gardner summarized the Lost Dutchman story and concluded that "a fabulously rich deposit in a form which enables it to be virtually had for the taking, probably does exist" (Gardner, p. 33). But aside from the "brooding menace" (Gardner, p. 33) of the Superstition Mountains, undocumented stories of Indian ambushes, earthquakes changing the area, and trigger-happy fortune seekers, what basis do we have for believing this story?

In the 1800s, the Peralta family apparently came north from Sonora, Mexico, to mine in the Superstition Mountains. They used Weaver's Needle as a landmark to locate their mine, but Indians ambushed them. Jacob Waltz, a Dutchman, rediscovered the Peralta Mine in the 1860s, and he confided the location of the rich mine on his deathbed around 1890. The story became public when newspapers ran accounts, with directions to Waltz's cabin in the western end of the Superstitions.

The books, *The Sterling Legend: The Facts Behind the Lost Dutchman Mine* (Gem Guides Book Co., California) and *The Lost Dutchman Mine of Jacob Waltz: Part I* and *Part II* (Cowboy Miner Productions, Arizona), are excellent sources for learning about the history of lore of this famous mine. It is important to note that in the Superstition Wilderness Area of the Tonto National Forest no roads or mechanized forms of transportation are allowed. In 1965 the area around Weaver's Needle was set aside as a public landmark and recreation area—no mining claims are permitted—and other lands within the wilderness area were similarly closed to mining after January 1, 1984.

A look at the geology of the area seems to argue against precious metal deposits being there. The recreated gold rush town of Goldfield near the beginning of the Apache Trail (Highway 88) reportedly had fifty short-lived mines in its vicinity in the 1890s, but it was well off the main body of the mountains. The Superstition Mountains are a huge volcanic field composed of tuff, ash flows, rhyolites, and lava flows from volcanic eruptions. Weaver's Needle is the resistant neck of a volcano. The Apache Trail apparently skirts the edge of several calderas as it winds through the mountains.

Under the thick deposits of barren rocks lies a basement of Precambrian granite where gold accumulated in patches as the granite weathered. The gold-bearing gravels forming the Whitetail Conglomerate are spotty in nature, and are not economic according to a study by Michael Sheridan and Sarah Prowell entitled, "Stratigraphy, Structure, and Gold Mineralization Related to Calderas in the Superstition Mountains" (*Arizona Geological Society Digest, Volume XVI*, 1986). But, as with many geological conclusions, even though the potential for a rich gold occurrence is considered to be small, it is still possible.

SEARCHING FOR THE LOST DUTCHMAN MINE

My interest was aroused by tales of the Lost Dutchman Mine, and I decided to revisit the area to check out Weaver's Needle. I picked up my *Apache Trail Treasure Map* from the Apache Junction Chamber of Commerce. It recommended a visit to the Lost Dutchman Museum at Goldfield Ghost Town. This museum is open seven days a week, except Christmas on State Route 88, four miles northeast of Apache Junction. The museum number is (602) 983-4888. Also available were jeep, helicopter, and horse-back riding tours of the Superstition Mountains.

I drove the Old West U.S. Highway 60 to the Peralta Road, midway between Apache Junction and Florence Junction. The exclusive community at Gold Canyon near the turnoff is nestled against the Superstitions, and other subdivisions are planned for the area. Plenty of camping spots were available off the washboard road, so I found a secluded place before sunset. Before it got dark, I climbed up a nearby hill to look over the country and I discovered rusty quartz veins crowning the summit. Since darkness was falling, I had to quickly make my way down to avoid getting skewered by the cactus and thorn bushes. Although I didn't bring back a sample of the ore, I felt confident it would make a good mine.

Early the next morning at the Peralta Trailhead, I paid my forest service day fee, loaded up my pack with water and lunch, and started hiking. Two old-timers were also starting out, and I joined them. Dick used a light aspen walking stick to steady himself on the rocky trail. I asked him if he knew about the legend.

"My great uncle on my grandfather's side was the Lost Dutchman," he matter-of-factly told me. "He told my grandfather things and my grandfather told me. I was just a boy, five years old."

We followed a dry streambed overgrown with brush. The shade was welcome, but the trail had boulders to skirt and false leads to figure out. We frequently rested, and sipped water to combat dehydration.

"He lived in town," Dick continued. "He was usually gone for a week at a time, so he must have spent three days going and three days returning. Twenty miles is quite a lot to walk in this heat. He may have ranged thirty miles from Apache Junction."

People believe the Lost Dutchman Mine in the Superstition Mountains of Arizona is located near Weaver's Needle. This wilderness area has popular recreation trails, and it is closed to mining.

When the trail left the creek bed and began to gain altitude along the canyon side, the men decided to turn back. "You'll run across all kinds of phony maps," Dick warned me. "If the mine is here, it could be anywhere."

I was sorry to lose their company. Dick might have known more, but he wasn't telling me. They said they were just taking a short hike to exercise, but I probably interrupted their surreptitious visit to the mine.

A family with heavy backpacks came down the trail and I looked them over. Perhaps they had some of that jewelry gold in their packs. The young sons seemed weary, but the father's eyes gleamed with triumph. I asked them where they'd been.

"First Water," the father answered. I recognized that as a famous location in the old accounts. His son took a swig from a gallon container of water.

"Did you find the Lost Dutchman Mine?" I boldly asked.

"Yup. We've got a load in the packs." The father grinned. "Can't wait to get back to the car at the trailhead." They swung off down the trail, and I was left in the brooding silence.

Farther up the trail I came across four women resting. One of them knew the legend, but the others played dumb. This was their third attempt to get a view of Weaver's Needle. Treasure seekers have a lot of determination, and they don't give up easily.

The trail switchbacked up the canyon for miles. No shade, and a blazing sun made me feel dizzy. I noticed a yellow layer in the strata across the canyon. No doubt that was pure gold hidden by a veneer of dust and sand. I didn't feel like leaving the trail just then, but I sketched the location on a scrap of paper.

When I finally reached the Fremont Saddle with its panoramic view of Weaver's Needle, I spotted two men up in the hoodoos on the ridge. They were obviously exploring uncharted country. I waited expectantly as they came toward me, searching their small fanny packs for signs of weight.

"Boy, that last stretch was difficult!" the tallest one exclaimed. He carried a water container strapped to his back. A plastic tube allowed him to drink without taking a break.

"Those rocks were a challenge," his paler companion agreed.

"What were you doing up in that rough country?" I pointedly asked.

The pale one pulled out a small topographic map. "We're doing cross-country hiking," he explained. "We started at the Carney Springs campground this morning at 6:00 a.m." His finger traced a route up a trail, and then along the contours of a ridge. "Not quite 2,000 feet of altitude," he boasted. An athletic young woman who was resting in the shade of the overlook came over.

"You guys doing cross-country?" she asked. "I'm in training for the Roosevelts."

"Now that's a tough climb," the pale hiker admitted. "You headed down? Let's talk about the best route." They seemed to sprint off into the bushes.

Several other hikers appeared as well as the four women I'd passed earlier. About a dozen of us found places in the shade and looked out at the view of Weaver's Needle. A jumbled mass of rocks guarded the slopes to the peak. By following the ridge from here, I could see a rough way to get closer. If I just had a better outfit, a week to explore, and there weren't so many prying eyes....

13. Fieldwork and Equipment

Ask a geologist about doing fieldwork and they will hesitate as they consider a few of the variables. Are we talking about Basin and Range where roads will get you close to a lot of places? Central American jungle where diseases may ruin a career? Subarctic muskeg and alder thickets?

"If you are driving the company truck," Bill Wilson of Alta Gold jokingly recommends, "carry jumper cables and park it pointed downhill." He's obviously talking about working in Nevada where some mining companies operate on a shoestring, and where geologists are used to driving close to a prospect.

"At the bottom of my pack I always carry a headnet," insists Sam Dashevsky, an experienced Alaskan geologist. "This past summer I was working out of McGrath, Alaska, and I lived under a headnet for three weeks."

"Everybody I know has a shoulder holster," recommends John Galey, Kinross Exploration's chief geologist for the Arctic region. "That way your arms are free. When I worked in Nevada, I had my handgun in my pack." Everyone has their own ideas about bear protection in Alaska. Having been salivated over by a big black bear, I prefer to lug a lightweight 30.06 rifle around. One man I worked for carried a shotgun with slugs even though we were next to a busy highway. One grizzly in his past had convinced him.

TERRAIN ANALYSIS

Exploration really begins with an analysis of the terrain and what is required for in-depth prospecting. Visit any exploration geologist's office and you'll notice that maps and map tables take up much of the space. Planning a strategy is an armchair activity using maps, aerial photos, and whatever written records are available, and it demands real analysis of how each unique terrain can be tested. In his *Introduction of Mineral Exploration*, Anthony M. Evans (Blackwell Science, London, England. 1995) calls this phase "desk studies" (Evans, p. 5)–the acquisition of publications and information about the selected area from government agencies and other sources. This includes geological and geophysical maps, USGS Mineral Resource Maps, USGS Open-file Reports, and geochemical surveys. Studying these types of data helps identify historic mining areas, geologic contacts, possible mineralized faults and veins, as well as access roads, cabins, and power lines.

Increasingly, the search for bonanzas requires the prospector to consider the application of new ideas about ore-forming processes, use a wide range of deposit models, and identify favorable tectonic features. Each succeeding generation of geologists has better tools and can reach deeper to find concealed ore bodies, and the prospector needs to utilize this new information to be effective. So background preparation begins with reading economic geology texts, taking classes, visiting known ore deposits and districts, and finding local resource people.

In his introduction to the *Mineral Prospecting Manual* (1987), a compilation from the French Bureau de Recherches Geologiques et Minieres, Jean-Bernard Chaussier states: "General mineral prospecting is essentially a natural experimental process." (Chaussier, p. 13) At each step in the process, the decision to continue or stop depends on the information gathered, the interpretation of that information, and how the hypotheses are modified. The manual begins with chapters on "preparations" (Chaussier, p. 13) and "the expedition" (Chaussier, p. 13) before going on to describe hammer prospecting, alluvial and eluvial prospecting, geochemical and geophysical methods, and drilling.

ORGANIZING YOUR SEARCH

During the initial phase, a prospector plots access routes, trails, and traverses. Usually, some clue has directed attention to the place, and there are questions to be answered about the source of minerals. Perhaps this is an undeveloped area with similarities to a productive locale. Edward Hargraves, an Australian who participated in the California Gold Rush, believed New South Wales had a similar potential for placer gold. Using techniques he had learned in California, Hargraves constructed a rocker, tested the creeks, and started Australia's Gold Rush. Since prospectors are independent, they don't have to justify every hunch they want to follow up, but there should be some compelling reason for selecting a particular area.

Most fieldwork involves gathering stream sediments, panning concentrates, sampling outcrops, and examining old workings. Assaying samples that are collected yield element amounts and possible anomalies that warrant follow-up work. Terrain, distance, access, ground cover, and other factors will limit how much work can be done, and a lean budget may mean fewer assays, or more emphasis on finding macro-gold or concentrating on surficial deposits. Like any good exploration geologist, a prospector needs to be thorough, covering the hard-to-reach places and exhausting all the possibilities before moving on.

Getting to know the country takes time and experience. Anthropologists suggest that indigenous people need ten years or longer to learn the resources and topography of a new area. I always plan on returning to a prospect at least ten times before enough information and samples are collected, and I find that topographic maps are invaluable for initial planning, and are essential on each outing. In addition, a fly-over with a small plane allows a good look at outcrops, old structures, trails, tailings, and altered rocks. I once located an old cabin hidden in the bushes during a Super Cub flight, and a mine dump next to the cabin yielded valuable information about the subsurface geology when I sampled it later.

IN THE FIELD

Expediting, that is, setting up a camp, arranging for replenishment of supplies, and getting out with a pile of samples are important considerations for any successful prospecting adventure. For years, I've selected targets based partly on accessibility by four-wheel-drive vehicle. All Terrain Vehicles (ATVs) are greatly extending our ability to get into remote areas. Although more expensive, airplanes and helicopters may also be necessary for accessing important targets, especially for the initial reconnaissance. Andy Harman, a Canadian mine-maker, explained how he set up prospecting camps when he took me out to a prospect in a helicopter.

"For a big project I would look for a good lake where I could fly in the supplies with a float plane and set up base camp. From there, helio time to get teams out into the brush was more affordable."

Four-wheelers can access remote areas and carry heavy samples out. Access is a key factor in prospecting.

He cajoled helicopter rental time from a mining company for the prospect we were going to investigate. Fortunately, the mining company that leased the helicopter saw the advantage of renting an hour of air time when they were not using the chopper. We stayed in the hills a week, staking ground and collecting stream and rock samples, and then Harman used a satellite phone to call in the helicopter a day early. From the air he plotted hunters' four-wheeler trails, which followed the ridges almost to the property. The next time he went in to the prospect, he had a team of four people drive in on four-wheelers. They stayed all summer, only coming out for supplies and for sending off samples.

"After the first week," Harman told me, "you get hardened in and really enjoy the camp life. I bring good meat and bury it in a frozen bank so it keeps. You gotta feed 'em well in the bush!"

On our initial trip, he took care to set the base camp on a breezy ridge that had water nearby. He cut boughs and made a comfortable bed; I slept on bumpy moss. We fried slices of a smoked ham for dinner, and had it again with eggs in the morning. There were easy-to-cook T-bone steaks and lamb chops for dinners, and he sautéed onions, mushrooms, and carrots for soups, adding left-overs and canned tomatoes. We ate well at the end of each hard day, making the trip feel like a catered safari. When I'm on my own, however, I usually just boil some noodle soup and add a tin of sardines for a quick meal.

The helicopter time might have cost $2,000, but because we were in totally untested country in the middle of a major mineral trend–it was worth it. Once on the ground, we learned where the game trails made hiking easier, laid out a claim block, and collected preliminary stream sediment and rock samples. Using this method of operation, Harman is able to identify and acquire any major mineral occurrences in the area.

NECESSARY ITEMS

What a prospector carries depends on the country they are prospecting. "As you get older, you have a checklist in the back of a notebook," advises John Galey, a geologist with the build of a football fullback. "Invariably, the first few weeks going into the field, you forget something."

He and other geologists use a surveyor's vest with a dozen different pockets for important items. At the end of the day it can be hung up, ready for use the next morning.

"A couple of marking pens, pencil and pen," he said, checking off the usual vest contents. "A hand pencil magnet, a roll or two of flagging, acid bottle with ten percent hydrochloric acid (in a plastic bag so it won't

John Galey, Kinross Gold's Arctic Region Chief Geologist, wears his multi-pocketed vest. Many of the important tools for fieldwork are kept in the vest.

accidentally leak and rot your clothes) for testing for carbonate rocks such as limestone or veins. A six-ring notebook with sample tickets, pocket altimeter, and a pair of small field glasses. When you are walking on a ridge or in bear country, it's nice to look over the outcrops from a distance. I also carry an extra compass, in case I lose one. The Silva compasses are good for brushing lines."

This reminded me of a time when I had gotten lost. Clouds had descended and I couldn't see familiar mountains to guide me. I was following a ridge and I got turned in a circle by using a spur off one side of the ridge. I eventually found my way home, but now I always figure out a compass bearing before I go off into the bushes. Even the most experienced geologists have stories about getting lost when the weather closed in or where the terrain was flat and featureless.

In Alaska, claim boundaries on state claims have to be north-south and east-west, so the orienteering compass is perfect for setting a straight course. Running a traverse is also easier if a compass is used. Many now prefer to use a Global Positioning System or GPS for field work, but getting exact bearing points for claim staking requires latitude and longitude fixes down to minutes and seconds.

"In my field pack," Galey continued, " I have a complete first aid kit, some heavy-duty seal-top plastic bags, sweater, wind pants, hat, gloves, candle, matches, mirror, BIC lighter, and waterproof matches in a water-tight container. I carry an extra hand lens and pencils in the pack so I don't get shut down if I lose any of them. I have rain gear all rolled up and ready—as soon as you don't think you need it, you do."

My rain gear is usually back in my pack when I'm off for a quick look at a nearby area. The clouds invariably move in. A small tarp helps in an emergency bivouac. I carry an orange one so it can be seen from the air.

"I have very little on my hips," Galey explained. "A Swiss Army knife and Swiss Army leatherman. I've always liked a twenty-four- or twenty-eight-ounce Stanley hammer, but these are a matter of personal preference," he admitted. "I have a long handled hammer for when I'm walking the ridges and breaking rock. And you gotta wear eye protectors. They're real eye-savers."

In the final analysis, prospecting is really about sampling, and Galey emphasized preparations for this activity. In a back pocket of his vest he carries a clipboard to hold topographic and geologic maps, and sample description sheets.

SAMPLER **Roger McPherson**

DATE	SAMPLE No.	LOCATION	DESCRIPTION	ALTER				PPM	
	RM008	Map:Circle B5-NE	Soil	BLCH	JSP		Au	0.01	
	R D Vn Vg / Ⓢ W SS	Upper Homestake		PRP	HM		Ag	0.4	pb/zn
		South Slope		ARG	JAR		As	138	
	Rsd	Below Rock Slide		AA	GOE	Sb	H̶g̶	2	220/216
	Tpd	UTM Zone:		PHY	CAR			14	
	Area	UTM N:		SIL	PY			<	
	Den	UTM E:		POT				<	
	RM009	Map:	Soil	BLCH	JSP		Au	0.01	
	R D Vn Vg / Ⓢ W SS	As Above		PRP	HM		Ag	2.0	
				ARG	JAR		As	116	
	Rsd			AA	GOE	Sb	H̶g̶	<	1540/131
	Tpd	UTM Zone:		PHY	CAR		Mo	592	
	Area	UTM N:		SIL	PY		W	<	
	Den	UTM E:		POT				<	
	RM010	Map:	Soil	BLCH	JSP		Au	0.01	
	R D Vn Vg / Ⓢ W SS	As Above		PRP	HM		Ag	0.2	
		Below Flag for		ARG	JAR		As	128	
	Rsd	Mag High		AA	GOE	Sb	H̶g̶	<	
	Tpd	UTM Zone:		PHY	CAR		Sb	22	98/130
	Area	UTM N:		SIL	PY		Bi	<	
	Den	UTM E:		POT			Mo	<	
	RM011	Map:	Schist–Quartzite with Milky Quartz Veins	BLCH	JSP		Au	0.02	
	Ⓡ D Vn Vg / S W SS		Some Carbonate	PRP	HM		Ag	0.2	
			Random Sample, Various Rock Types	ARG	JAR		As	156	
	Rsd		Spotty FeOx	AA	GOE		H̶g̶	<	
	Tpd	UTM Zone:		PHY	CAR		Sb	10	18/40
	Area	UTM N:		SIL	PY		Bi	<	
	Den	UTM E:		POT			Mo	<	
	RM012	Map:	Schist–Occasional Quartz Vein, Moderate FeOx	BLCH	JSP		Au	0.03	
	Ⓡ D Vn Vg / S W SS	Same Area	Random Sample of Float	PRP	HM		Ag	18.0	
		as RM009		ARG	JAR		As	1570	W-10
	Rsd			AA	GOE		Hg	<	pb/zn
	Tpd	UTM Zone:		PHY	CAR		Sb	7930	>10,000/58
	Area	UTM N:		SIL	PY		Bi	<	Cu-468
	Den	UTM E:		POT			Mo	<	Cd>100
	RM013	Map:	Soil	BLCH	JSP		Au	0.02	
	R D Vn Vg / Ⓢ W SS	Upper-Middle		PRP	HM		Ag	<	
		Homestake Ck.		ARG	JAR		As	14	
	Rsd	North Slope		AA	GOE		Hg	<	
	Tpd	UTM Zone:		PHY	CAR		Sb	36	68/80
	Area	UTM N:		SIL	PY			<	
	Den	UTM E:		POT				<	
	RM014	Map:	Soil	BLCH	JSP		Au	<	
	R D Vn Vg / Ⓢ W SS	Upper-Middle		PRP	HM		Ag	<	
		Homestake Ck.		ARG	JAR		As	16	
	Rsd	North Slope		AA	GOE		Hg	<	112/100
	Tpd	UTM Zone:		PHY	CAR		Sb	56	
	Area	UTM N:		SIL	PY			<	
	Den	UTM E:		POT				<	

FORM FM_SM002

A filled-out form example of sample descriptions. Describing a sample and its location is easier when using a standard form. Rock samples from the author's claims show some mineralization. Assay results were added later.

"Do everything systematically," Galey's voice became insistent. "Where the samples were collected, a good description–the more information, the better. Get a big enough sample, a statistical representation of the outcrop, and write a general description of the rock, noting any veining or alteration. Keep a hand sample. Use a good assay lab that will provide reliable numbers. How you assemble that information will inevitably help guide you to a deposit. If all your data is systematically collected, it keeps the quality to a high level of consistency."

Galey gave me sheets for describing each sample I collected. I've found these forms force me to look at each sample more closely. A typical entry asks for: the date; an identifying number; whether it's rock, vein, soil or stream sediment; its location; a description; any alteration; and a place for noting the assay results when they come in. On the reverse side of the sheet is space for sketching the rock and the outcrop, and mapping its location.

"Sampling is as much an art as a science," Galey believed. "If you're looking for gold, you've got to give yourself the best opportunity for finding gold in that sample. Get a sense of how big the outcrop is. Make a sketch map. Any veins? Where might the gold be? What are we gonna sample? If you don't see a vein, take a lot of pieces at a statistical spacing over that area. Get a whole lot of pieces, enough to fill a bag with five or six pounds. If there's a vein, sample the host rock separately from the vein. Then high grade the heck out of the vein to see if there's any gold!"

For a mineralized area where there's a lot of outcrop, Galey recommended treating the surface as if you were underground in a mine. "With a clipboard, sketch it to scale by careful measuring, and use a compass to plot it on a map."

For soil samples, use a shovel to find out if there is a developed soil horizon. "If you are doing a soil sampling program," Galey added, "you can get a screen to sieve out coarse pebbles and organics. You want five hundred grams [about four cups] of soil sample. You could use those zip-lock freezer bags. They have a place to write on and they're light."

WHERE TO FIND EQUIPMENT

Mining supply catalogs carry field equipment, hand tools, notebooks, compasses, and sample bags. Cotton or polyethylene fiber bags about seven by twelve inches with a yellow identification tag and draw strings are fairly standard for rock samples; soil sample bags can be smaller.

Soil sampling bags three by six inches of heavy paper or plastic are also used–as long as identifying numbers can be written on the outside. Write to Miners Incorporated, P.O. Box 1301, Riggins, ID 83549-1301

TIPS ON WHAT TO BRING

John Sinkankas' *Prospecting for Gemstones and Minerals* (Van Nostrand Reinhold Co., New York. 1970) has chapters on using maps, camping, clothing, field notes, and rock identification. Also, there is a chapter on tools that mainly focuses on collecting and preserving crystallized specimens. Under camping tips, Sinkankas recommends a camp-out kit with durable utensils: a can opener, plates, cups, and a standard check-off list for food items. Fred Rynerson's *Exploring and Mining Gems and Gold in the West* (Naturegraph Company, Healdsburg, California. 1967) is a classic account of desert prospecting by horse and by car in the early 1900s when the jeep was starting to replace the burro.

Adventuring in the California Desert by Lynne Foster (Sierra Club Books, San Francisco, California. 1987) covers basic gear, desert survival, and camping. Foster's "ten essentials" (Foster, p. 81-98) are: maps, compass, flashlight, sunglasses, extra food and water, extra clothing, waterproof matches, candle or fuel tablets, pocket knife, and first aid kit. Toilet paper is the eleventh essential.

Those who are only prospecting for placer gold, you will need: a gold pan; screens for sifting and classifying the material to reject larger pebbles and rocks; a pointed shovel; pry bars; a hammer; and crevicing tools–pointed gads, brushes, and dental tools. Magnifying glass, plastic vials, maps and compass are also useful. A lightweight portable sluice box and five-gallon plastic bucket should be carried for processing larger quantities of material. Panning for gold also yields heavy minerals, which naturally lead on to a search for lode sources. Once in the field, a prospector has to be open to recognizing clues to many types of mineral deposits.

MODERN PROSPECTING VS. THE LONE PROSPECTOR

A textbook on the subject, *Exploration and Mining Geology* by William C. Peters (John Wiley and Sons, NY. 1978) characterizes early exploration as "prospect examination" (Peters, p. 13) until the mid-1900s. Typically, a prospector would report a discovery, and a geologist or mining engineer would assess the extent of mineralization. Peters feels that the opportunities for the independent prospector have diminished as near-surface high-grade ore bodies have been found and mined.

Recent uranium and gold prospecting were the last examples of "one-person, one-objective" fieldwork. Emphasis has changed to district-wide and regional efforts with favorable regions, areas, and targets being identified for thorough examination. This view sees prospecting as a group effort by trained geologists searching for new ore districts. "Geophysics, geochemistry, and drilling technology provide the team with tools of discovery," Peters writes. "Laboratory and the computer facilities provide the tools of measurement."

"Large-company helicopter-planned traverses cover a lot of ground," geologist Sam Dashevsky conceded. "But they insulate us from miles of ground we're not walking on. The helicopters get us into new areas, but you've got to have crews to cover the hard terrain and get dirty to find out what's under there."

In my view the individual prospector has many advantages the mining company lacks. Flexible, highly motivated by personal interest, working on a very low budget without corporate constraints, and supplied with the same basic tools as the trained geologist, the individual prospector determines for him- or herself what to emphasize and where to go. If a prospector has insights and discovers mineralized prospects, a mining company may become interested. As long as the fieldwork is done carefully, the sample results duplicatable, and the data presented in a readable report, the discovery will be explored more thoroughly.

Consider the situation facing a new mine with a life of eight to ten years. The capital costs for the mill and processing facility are paid back by mining the ore body. However, if similar deposits are found within a radius of fifty or so miles, the mine benefits from a new supply of ore, and the capital costs are amortized even more.

Meanwhile, the mine has geologists out combing the hills for new deposits, but their budget is limited and prospect drilling eats up a big part of the exploration funds. An independent prospector working in this district may find new targets that the mining company is willing to investigate. Sometimes after an initial data submission, a large company will pay for a prospector's assays. This has happened to me, and it helps both by reducing my expenses and by having my assays processed more quickly because they are submitted by a large company.

PHYSICAL AND MENTAL PREPARATION

In addition to finding new places to prospect and remembering to bring all the essentials, a person needs to be in decent physical shape. In winter, I walk an hour every other day, and I spend an hour working up a sweat in a gym with weights and aerobic equipment on days I'm not walking. I like prospecting because it gets me outdoors and is physically challenging. I voluntarily climb mountains to look over the rocks and investigate brushy and hard-to-reach places. There are moments in winter when I'm dripping sweat from a Stairmaster workout, and I recall this same feeling from hiking uphill with a pack full of samples. It reminds me that my body is the most important part of my prospecting equipment.

Prospectors must be prepared mentally and physically, and have the right instruments. Identifying a major trend, interpreting favorable tectonic features, and extending the search out from known deposits and districts are all part of modern prospecting. It's a challenge to find new areas to explore, and to figure out how to achieve the best coverage of the ground. This is a creative process that analyzes how to really explore a terrain, and how to go beyond previous investigations. Just don't forget the rock hammer, sample bags, and water bottle!

14. Modern Assaying

For about $20 to $30 a sample a person can get an assay of over thirty elements at the parts per million level, and gold in parts per billion. A two-pound chunk of rock, the concentrates from three gold pans, a handful of stream sediments, or a bag of soil can be assayed. Keep in mind that assays are to detect microscopic amounts of gold and other elements during the exploration phase of prospecting; visible gold in a gold pan or rock sample will result in a meaningless assay since you already know gold is present.

Assaying can provide critical answers when doing regional reconnaissance surveys, finding specific targets within a known mineral district, identifying cut-off grades to mine, or evaluating the mineral values in a vein or a disseminated deposit. Knowing the limitations of each analytical method is important when selecting an assay package because the price of a multi-element geochemical assay will vary depending on the number of elements assayed and their detection levels. Many laboratories don't like to handle less than 20 samples, so a person should have a serious prospect under study rather than just be curious about a rock.

A TOUR OF ALS CHEMEX

ALS Chemex (www.alschemex.com), one of the larger providers of assay and geochemical services to the North American and international mining and exploration industry, has several multi-element trace level Inductive Coupled Plasma (ICP) assays that are suitable for prospecting reconnaissance programs. Costing from $10 to $20, excluding preparation charges, thirty to fifty elements are measured. Trace level detection of gold in the one to five parts per billion range using a fire assay process costs an additional $11.

"The typical analysis was gold at five parts per billion and 34 element ICP—but we keep adding more elements to the package," Lawrence Ng, Preparation Manager for the Chemex Vancouver facility told me.

The Chemex Company was founded in Vancouver, Canada in 1966 and it merged with ALS in 1999. The company now has labs in Nevada, Alaska, Montana, Canada, Mexico, South America, Australia, Asia and other countries. Their fee schedule can be obtained by writing to:

CHEMEX
994 Glendale Avenue, Unit 7
Sparks, Nevada 89431
www.alschemex.com

The tile-roofed ALS Chemex prep and fire assay lab in Sparks, Nevada has an appropriate rock motif on its tree-shaded exterior. As I was getting out of my car, a young man was struggling to unload large, white sacks from his pickup truck. A receptionist came running out.

"If those are samples, hon," she said, "You can drive around back and unload right onto the pallets."

Behind the quiet offices was a warehouse area where stacks of samples were waiting to be processed. I was given a tour starting from where the samples come in and are identified on a computer worksheet, past the walk-in drying oven and rock crushers (which are made of chromium steel–this adds some chrome to the assay when harder material is crushed), through a soils preparation area (one to two pounds of soil is ideal). From there the samples are sent on to Vancouver for assaying. Fire assaying for gold and silver is done in Sparks.

We entered the fire assay lab, where they do gold and silver analyses, through a door with skull and crossbones and the message, "Abandon all hope ye who enter here." Because lead is used in the process, it's a potentially poisonous work area. They have two fusion furnaces that heat up as many as 84 individual samples to 2300°F at a time. Trays of small cupels with beads of gold and silver were cooling on the table. In the atomic absorption lab a fellow in a white lab coat and light blue rubber gloves showed me how he obtained a measurement for gold.

"It's all electronics and computers," the technician said. "Once it reaches equilibrium, and the number doesn't change, the computer takes the data and does the rest." To insure quality control, a number of assays within a batch are redone to see if the numbers turn out the same.

Geologist	
Phone	
Project	# of samples
P.O. #	
Date	Quote #

Chemex Labs

212 Brooksbank Avenue
North Vancouver, BC, V7J 2C1
Tel: 604-984-0221 Fax: 604-984-0218

994 Glendale Avenue, Unit 7
Sparks, NV, 89431
Tel: 702-356-5395 Fax: 702-355-0179

Sample type code

☐ S Soil	☐ A Pan concentrates
☐ L Sediment	☐ H Heavy minerals conc.
☐ R Rock	☐ C Mill concentrates
☐ D Drill	☐ V Vegetation
☐ P Perc. cuttings	☐ W Water
☐ X Pulp	☐ O Other

RESULTS DISTRIBUTION

-Mail original to-	-Mail copy to-	-Fax to-
		Fax #:
		Attn:
		Fax #:
Attn:	Attn:	Attn:

INVOICE DISTRIBUTION   **Additional Instructions**

-Mail- ☐ to same as original results
　　　☐ to:

Rush services available upon request at 1.5 x list price (see reverse)

Attn:　　　　　　Rush parameters:

Please, Specify sample type code or prep code as necessary SAMPLE NUMBER (16 characters maximum)	TRACE	ASSAY	Au Code	Ag	As	Sb	Hg	Cu	Pb	Zn	TRACE		ICP		WHOLE ROCK	
											G4	G7	G32	G24		

IF SAMPLES ARE TO BE ANALYZED FOR A COMBINATION OF TRACE AND ASSAY PARAMETERS, PLEASE USE THE LETTERS T AND A INSTEAD OF CHECK MARKS

PLEASE, SEE REVERSE SIDE FOR CHEMEX POLICY ON STORAGE OF PULPS AND COARSE REJECTS

Mississauga, Ontario
5175 Timberlea Blvd., Mississauga, ON, L4W 2S3
Tel: 905-624-2806 Fax: 905-624-6163
Thunder Bay, Ontario
920 Commerce St., Unit 5, Thunder Bay, ON, P7E 6E9
Tel: 807-475-3329 Fax: 807-475-9196
Rouyn, Quebec
175 Industriel, CP 284, Rouyn, PQ, J9X 5C3
Tel: 819-797-1922 Fax: 819-797-0106

Elko, Nevada
651 River Street, Elko, NV, 89801
Tel: 702-738-2054 Fax: 702-738-1728
Butte, Montana
103 North Parkmont Industrial Park, Butte, MT, 59701
Tel: 406-494-3633 Fax: 406-494-3721
Tucson, Arizona
2015 North Forbes Blvd., Tucson, AZ, 85745
Tel: 520-798-3818 Fax: 520-882-2030
Anchorage, Alaska
5640 B Street, Anchorage, AK, 99518
Tel: 907-562-5601 Fax: 907-562-6502

Hermosillo, Mexico
Periferico Poniente No. 144 entre Blvd. Navarette y Blvd.
L.Encinas Colonia Raquet Club, Hermosilla, Sonora, 83200
Tel: 62-604-475 Fax: 62-604-476
Guadalajara, Mexico
Jazmin 1132, entre R. Michel y Amapola, Sector Reforma,
Colonia San Carlos, Guadalajara, Jalisco, 44460
Tel: 3-619-4616 Fax: 3-619-4616
Zacatecas, Mexico
Avenida H. Colegio Militar No. 237, esquina Calle Lomas de la
Estacion, Colonia Sierra de Alica, Zacatecas, Zacatecas, 98050
Tel: 492-2-99-88 Fax: 492-2-99-88

White and yellow copies - include with shipment　　　　　　*Page ____ of ____*

The assay laboratory order form tells the lab what kinds of samples are enclosed (rock, soil, stream sediment, or pan concentrates), which geochemical assay package to use (gold plus thirty-five elements is typical), and what to do with the extra material and assay pulps (storage for a fee or discard). Assay labs have booklets describing assay packages and listing charges.

ACME ANALYTICAL LABORATORIES LTD.
REQUISITION FOR ANALYTICAL WORK
852 East Hastings St. • Vancouver, BC • V6A 1R6 • CANADA • E-mail: info@acmelab.com • Tel: (604) 253-3158 • Fax: (604) 253-1716 • Toll Free: 1-800-990-2263

Carrier:	# of Parcels:	Req. Number:	*Acme File Number*
Waybill:	Date Received		

CLIENT INFORMATION

CLIENT: ☐ Certificate ☐ Invoice	COPY TO: ☐ Certificate ☐ Invoice
Company:	Company:
Address:	Address:
Attn:	Attn:
Phone: Fax:	Phone: Fax:
Submitted By: PO #:	Project: Date: / /

WORK PRIORITY ☐ Regular ☐ **RUSH!** NEEDED BY: / /

DATA TRANSFER ☐ Diskette ☐ Modem ☐ Fax:_____ ☐ E-mail:_____

ANALYSES REQUIRED

Type of Sample	Number of Samples	Sample Sequence From - To	Prep Code	Analytical Package or Elements Wanted	Remarks (ie. Specify package options)
		-			
		-			
		-			
		-			
		-			
		-			
		-			

COMMENTS

STORAGE

Please Note: Acme provides free storage for a period of 1year for all sample pulps and 3 months for rock or drill core rejects. Soil rejects are discarded immediately unless otherwise directed by the client.

Rejects: ☐ Discard ☐ Return ☐ Store for _____ yrs

Pulps: ☐ Discard ☐ Return ☐ Store for _____ yrs

Return To: Company:_____ Address:_____ Attn:_____ Tel:_____

FOR YOUR REFERENCE

Preparation Methods

SS80 - Soil and Sediment Prep. (-80 Mesh)
VA80 - Ash vegetation at 475°C
R150 - Rock and Drill Core Prep. (-150 Mesh)
R200 - Rock and Drill Core Prep. (-200 Mesh)
M150 - Metallics Prep. (-150 Mesh)
M200 - Metallics Prep. (-200 Mesh)
P150 - Pulverize only (-150 mesh)
P200 - Pulverize only (-200 mesh)
MPMP - Mixing and Pulverizing Composite
PSCB - Pulverize by Ceramic Mill
HMC1 - Heavy Mineral Concentrate (2.96 SG)
HMC2 - Heavy Mineral Concentrate (3.32 SG)

Geochemical Analysis & Assay Methods

Group 1C - Hg by Cold Vapour AA
Group 1D - 30 Element ICP Analysis - Aqua Regia
Group 1DX - 35 Element Optima ICP Analysis - Aqua Regia
Group 1E - 35 Element ICP Analysis - Total Digestion
Group 1EX - 40 Element Optima ICP - Total Digestion
Group 1F-MS - Ultratrace by ICP-MS (Specify: 1, 15 or 30 gm)
Group 1T-MS - Ultratrace by ICP-MS - Total Digestion
Group 1TIL - Multi-element Cyanide Leach
Group 1ENZ - Enzyme Leach ICP-MS Analysis
Group 1SL - Sequential Leaches by ICP-MS (specify type of leach)
Group 2A - Special Extraction or Carbon & Sulphur
Group 2B - Water Analyses
Group 2C - Water Analyses by ICP or ICP MS

Group 3A - Au by Wet Extraction
Group 3B - Fire Geochem for Au ± Pt, Pd, Rh
Group 3C - Au by Cyanide Geochem Leach
Group 4A - Whole Rock by ICP
Group 4B - Whole Rock Trace Elements by ICP-MS
Group 4X - Whole Rock by XRF
Group 5A - Neutron Activation Analysis

ASSAYS

Group 3D - Au by Cyanide Assay Leach
Group 6 - Precious Metal Assay
Group 7 - 15 Element Assay by ICP
Group 8 - Single Element Assay

Special Exploration Packages

Geochemical Analyses

Geo 1 - Group 1D + Group 3A
Geo 2 - Group 1DX + Group 3A
Geo 3 - Group 1C + Group 1D + Group 3A
Geo 4 - Group 1D + Group 3B for Au, Pt, Pd
Geo 5 - Group 1C + Group 1DX
Geo 6 - Group 1EX + Group 3A

Assay Packages

Assay 1 - Group 7 + Group 8 Au Wet Assay
Assay 2 - Group 7 + Group 6 for Au, Ag
Assay 3 - Group 6 for Au + Group 1DX

form: 1003a.xls 10/25/99

A sample Acme Analytical Laboratory order form.

PATHFINDER ELEMENTS

When a prospector receives an assay print-out, interpreting the results requires some knowledge of the elemental associations in nature. The gold-associated pathfinder elements arsenic, antimony, and mercury are usually on the list because they disperse widely around a deposit. Lead, zinc, copper and silver are also part of a multi-element assay package. A few lanthanides (that group of elements isolated in a separate strip in the Periodic Table of Elements) such as cerium, scandium, and lanthanum are tested for, in case it's a rare earth deposit. Lithium or beryllium and fluorine would be pegmatite indicators. Barium, cadmium, and manganese suggest volcanogenic (undersea, rift-related copper-lead-zinc) deposits. Aluminum, potassium, sodium, rubidium, strontium, and zirconium amounts indicate different kinds of igneous rocks. Thallium tends to occur in hydrothermally altered rocks that are associated with ore deposits, bismuth in granite is considered a good indicator of gold potential. Tungsten (W) points to a high temperature deposit related to magma. High nickel and iridium could point to platinum metals.

In 1974 Robert W. Boyle wrote about these relationships for the Geological Survey of Canada *Paper 74-45*: "Elemental Associations in Mineral Deposits and Indicator Elements of Interest in Geochemical Prospecting." In his introduction he states: "Concentrations of a single element rarely occur in the earth. More generally a suite of elements is concentrated in a particular deposit because of certain intrinsic chemical properties which depend essentially on their electronic constitution and hence their position in the periodic table....[C]ertain elements may serve as indicators of others that are too low in abundance....Thus arsenic in rocks, soils, and vegetation may indicate the presence of gold deposits, and nickel may be indicative of platinum metals in certain terranes" (Boyle, p.1-2).

Knowing what constitutes an anomalous amount above the normal crustal abundance helps when reading an assay print-out. Tables of background values of elements are published in geochemical exploration texts, although these values vary considerably because of natural and chemical processes. Establishing background values for each prospect area requires some sampling of the surrounding unmineralized rocks and soils. In addition, regional anomalies may mask local targets, giving a high background level because of widespread enrichment.

BONDAR-CLEGG

Geochemical Lab Report

Inchcape
Testing
Services

DATE PRINTED: 17-SEP-93
REPORT: V93-00871.0 (COMPLETE)   PROJECT: NONE GIVEN   PAGE 1

SAMPLE NUMBER	ELEMENT UNITS	Au PPB	Cu PPM	Zn PPM	Ag PPM	As PPM	Sb PPM	W PPM
S4 2H-2		36	28	57	<0.1	277	16.0	3
S4 2H-3		230	33	62	<0.1	159	7.4	4
S4 2H-4		68	34	60	<0.1	187	7.7	3
S4 2H-5A		55	26	60	<0.1	220	20.7	6
S4 2H-5B		39	36	63	<0.1	189	5.5	<2
S4 2H-6		130	28	52	<0.1	236	12.0	7
S4 2H-7		27	29	59	<0.1	109	13.0	4
S4 2H-8		36	32	66	<0.1	138	13.0	5
S4 2H-9		21	31	63	<0.1	112	8.1	3
S4 2H-10		42	31	66	<0.1	343	11.0	3
S4 2H-11		22	34	75	<0.1	77	4.8	<2
S4 2H-12		12	30	69	<0.1	80	5.0	<2
S4 3H-0		15	31	64	<0.1	100	4.4	3
S4 3H-1		17	24	63	<0.1	59	4.7	4
S4 3H-2		10	20	54	<0.1	56	6.2	<2
S4 3H-3		14	23	58	<0.1	73	7.0	<2
S4 3H-4		15	23	64	0.1	95	11.0	3
S4 3H-5		13	25	59	0.2	67	13.0	<2
S4 2J-1		32	17	42	<0.1	237	14.0	<2
S4 2J-2		140	17	40	<0.1	249	17.0	<2
S4 2J-3		62	21	42	<0.1	167	19.0	<2
S4 2J-4		50	28	52	<0.1	221	14.0	3
S4 2J-5		52	32	54	<0.1	211	6.0	3
S4 2J-6		9	35	61	<0.1	90	3.5	<2
S4 2J-7		8	35	59	<0.1	97	5.0	3
S4 2J-8		9	34	63	<0.1	45	3.1	<2
S4 2J-9		9	30	64	<0.1	68	6.3	<2
S4 2J-10		19	29	54	<0.1	88	4.9	2
S4 2J-11		25	30	60	<0.1	147	4.4	<2
S4 2J-12		12	30	65	<0.1	65	5.4	<2
S4 3J-0		7	26	61	<0.1	46	3.9	<2
S4 3J-1		9	24	58	<0.1	62	5.8	<2
S4 3J-2		22	28	67	<0.1	159	8.5	<2
S4 3J-3		<5	25	58	<0.1	50	4.6	<2
S4 3J-4		<5	28	63	<0.1	35	3.6	<2
S4 3J-5		17	28	63	<0.1	74	4.9	<2
S4 3J-6		9	23	59	<0.1	70	4.5	3

An assay print-out of a line of soil samples lists the sample numbers, and gives the element abundance in each sample. Anomalies above background should stand out. Pathfinder elements such as arsenic disperse more widely in soils, helping target areas for follow-up testing.

Care must also be taken when collecting samples to avoid contamination. Drill cuttings may have higher tungsten or chromium levels because of metal that is ground off the drill bit. Collecting or storing samples in galvanized buckets will add higher zinc values. In addition, some results may show a single high from a "nugget effect," a chance occurrence of gold that skews the results. One line of soils I had assayed contained a sample with 1,500 parts per billion gold while the others were around 100 parts per billion gold. Since I couldn't duplicate this high value during additional sampling over the area, I concluded that it was a chance "nugget."

GEOCHEMICAL EXPLORATION

Geochemical exploration for mineral deposits originated in the U.S.S.R. and Scandinavia in the 1930s and spread to the U.S. by the 1950s. The development of atomic absorption spectronomy (AA) in the late 1950s allowed the accurate analysis of many elements. More recent assay techniques, using inductively-coupled plasma (ICP), have speeded up the process and made it more affordable. In an ICP analysis, the processed sample passes through a high energy plasma and the elements emit light at their characteristic wavelengths. This, and other processes, are explained in assay laboratory brochures, along with the limitations of each analysis. The SGS (Soceiete Generale de Surveillance) Group publishes an excellent booklet that explains the assay methods and has a glossary covering terms such as ICP, AAS, and INAA. An SGS thirty-two element package using multi-acid digestion costs about $13.50, excluding sample preparation (drying, crushing, sieving, etc.), and a fire assay for gold would add about $12. See their web site at www.SGS.com/minerals.

An assay reveals the mineral content of rocks, soils, or a stream drainage which may not be evident even under a microscope. Frequently, when I'm bent over washing gravel from my gold pan, people assume that I'm panning for gold. "You're a prospector? I bet you've seen a lot of nuggets." Dealing with parts per million means that I often don't see minerals in my gold pan. What I'm doing is concentrating the trace elements that have come from a large drainage area to produce pan concentrates for assaying.

Once, when I found an old prospect dump near an unmined creek, I took a twenty-pound sample to the nearest water and panned it down to get concentrates. An assay told me that there was a gold placer the old-timers hadn't exploited. The assays won't tell me how rich the placer is; I would have to trench or drill the creek to bedrock and do careful panning and sampling to find that out. But at least I have solid information for claiming the area.

Rocks are tested for leakage halos and diffusion aureoles from "blind orebodies," and for tracing zoning and direction of an orebody. Nevada geologists have told me: "Assay all the rocks." They've learned that microscopic gold is undetectable except through assaying. However, doing a lot of rock assays is too expensive for most people, so some limitations are necessary. Veined rocks, altered rocks, colorful rocks, iron-stained rocks–these are all likely candidates for assaying. In some areas of Nevada, impure limestones and jasperoid would be targets.

Just any rock from a creek bed is not likely to be important. The rocks must be from a particular bedrock location that can be found again. Stories of rich jewelry rocks are often followed by fruitless searches for their origin. I make a sketch map of where each sample came from and write a description of the sample. When I send in rocks for assaying, I keep a representative piece so I will know exactly which one had the good values.

The real growth in mineral exploration has occurred in soil sampling. When a target has been selected (perhaps because of an old gold placer, historic mining district, or mineralized float in a creek), soil sampling is done at intervals of between 100 and 400 feet. When the ground isn't too rocky, I use a small shovel to quickly fill up a five inches by seven inches cloth sample bag–picking out the rocks, or course. Exploration companies use gas-powered augers with a two- to three-inch diameter borer to cover the ground quickly. In rocky ground, collecting a good soil sample can be strenuous work.

The principle behind soil sampling is that mineral deposits have a halo around them of more mobile pathfinder elements, such as arsenic and zinc, which disperse widely in altered rocks, soils, and stream sediments. Mercury and antimony (stibnite) can also indicate a potential precious metal deposit. Soil sampling becomes an important exploration tool once an area has indications of mineralization. Sample lines are laid out along slope contours to find downslope dispersion fans from a hidden deposit. Soil horizon depth varies depending on each area's soil formation and bedrock. The "B" horizon is considered the best depth to sample–this is below the "A" horizon, or organic layer, and above the "C" horizon of decomposing bedrock.

Once samples are collected, packaged in cotton bags, plastic bags, or cardboard envelopes, and given identifying numbers, they are ready for sending in. Assay labs have submittal forms with a choice of geochemical or ore grade analysis, columns for sample type (rock, soil, pan concentrates, or stream sediment), sample number (I make up a number such as 05-1A and keep a description of what each number represents), and a choice of their multi-element packages (Gold Plus Thirty-five Elements is typical).

Coarse rejects, the part of the sample not analyzed, are generally discarded by the lab after several months unless otherwise requested, and the pulps from the analysis can be either saved for a future analysis (for a storage fee) or discarded. To avoid additional charges for storage, an order should state how coarse rejects and pulps are to be handled. Order forms usually include this question.

PREPARATION LABS

Many of the larger assay labs have "prep" labs near active mining areas. Rocks are crushed to a flour-like fineness; soils and stream sediments are dried and screened to negative eighty-mesh; and pan concentrates are dried and screened. Then the prepared samples are sent on to the nearest assay lab. Preparation of each sample costs approximately $3 for soils, $6 for pan concentrates, and $5 for rocks. Sample bags can be purchased from a catalog which is put out by Miners, Inc., Riggins, Idaho 83549, or from Legend, Inc., 988 Packer Way, Sparks, Nevada 89431.

A few blocks from the ALS Chemex lab in the same industrial area of Sparks, Nevada, I visited the modern prep lab of American Assay Laboratories. Lab Manager Robert Annan, looking like a football fullback in a business suit, showed me around the operation. Like many in the business, he was a chemist with twenty-five years experience.

"When I got here three years ago, gold prices were just picking up," he said. "The last two years have been really good." We skirted around two hot fire assay furnaces. "We have a thirty-four element package that will run $11, plus prep work will run about $14. That will get you a wide range of elements for exploration. That should be good for a small prospector." For more information go to www.aallabs.com or write American Assay Laboratories, 1599 Glendale Avenue, Sparks, Nevada 89431.

Finding a good prep lab is the first step in obtaining an assay. Since I work out of Fairbanks, Alaska, I've used both Chemex and Alaska Assay Laboratory. Dave Francisco of Alaska Assay chooses from several assay labs in order to give his customers the best package.

"I see both sides of the story," he told me. "About half my career has been in commercial business and half out at the mine site. I understand not only the exploration side, but also what the mine labs are doing, what the metallurgy group is doing, and what the environmental groups are doing."

Behind him an aerial photo of the enormous Bullfrog pit mine near Death Valley served as a reference to his past work experience. I could hear the muffled roar of the lab dust collection system, and the deeper thrum of the crushers as we talked. Workers popped into the office with questions, cranking up the noise level each time the door opened.

"There's three stages in exploration assays," Francisco informed me. "There's the actual taking of the sample at the drill or auger or in the field by hand. Then you've got your sample preparation procedure which is one of the most important because if you don't do that properly there's no use continuing on with the analytical procedure. The samples are brought in, dried, crushed, prepared down to a sub-sample. Those samples are then able to be shipped less expensively down to the Lower 48 or to Vancouver for analysis. It's like a pyramid effect with the variances. It starts with the sampling, then you've got the sample prep, and then you've got the analytical variances.

Having had a lab lose some of my samples and then depend on stored splits that returned false results, I really understood Dave's concern about procedures.

His lab sees all varieties of reconnaissance samples: soils, stream sediments, pan concentrates, rocks and drill cuttings and core. The lab dries everything, but samples that come in plastic instead of cloth bags have to be transferred to pans for drying. "I think it's much easier to put samples in a cloth bag," he advised. "It's easier for us to put the reject material back in the bag as well."

When I brought in soil samples for assay, he recommended a thirty element ICP package and a fire assay for gold detection at 5 parts per billion.

To view the complete process of assaying, I visited the Acme Analytical Laboratories in Vancouver, British Columbia. Chief Assayer Clarence Leong gave me a quick tour, beginning with the dock where samples come in from worldwide locations, usually already in powder form. A cardboard box filled with numbered two by three inch brown envelopes was a typical sample submission. Leong gestured toward some large bags on a cart.

"It's rare that they come in like this," he remarked. "We'll crush the sample, and a split of 250 grams will be pulverized to 150-mesh before analysis."

The downstairs was used for drying, pulverizing, and screening the samples, and assigning a tracking number and worksheet. Upstairs, the samples were fire assayed for gold, and others sent through ICP for a multi-element analysis. I watched as test tubes containing samples to assay were moved through various procedures. "This is the digestion area," Leong explained. "Most [gold samples] are 'digested' in aqua-regia in a hot water bath for one hour."

After digestion, or dissolving by acids, the test tubes were sent to another station for analysis. I watched a rack of forty test tubes with yellow liquid at the ICP machine. An automatic probe dipped into one tube, then washed itself before going on to the next tube in the batch. "The thirty element ICP analysis is popular for a lot of rock and soil geochemistry," Leong commented. "The new generation of ICP equipment will give us a lot more elements to analyze. It's more sensitive, also."

When we toured the fire assay area, with its ceramic cupels and furnaces, Leong took time to explain the difference between fire assay and geochemical assays.

"Fire assaying is for ore-grade and geochemical analysis for precious metals. Assay data is critical [information] because a company uses this data for ore estimations. Geochemical analysis is designed to help reconnaissance programs determine where the ore body is." Contact the lab for a price brochure at:

ACME ANALYTICAL LABORATORIES
852 East Hastings Street
Vancouver, British Columbia
V6A 1R6
Canada
www.acmelab.com

SUMMARY

Once thought to be a mysterious and complicated process, assaying is an inexpensive and important tool for prospecting. While mining companies send in thousands of samples for assaying each season, individuals like myself also send in small quantities for assaying. Getting an assay print-out can justify the start of a mine or signal the end of a prospect. But without such hard data, the pretty rocks will remain just a curiosity on the shelf.

15. GEOCHEMICAL PROSPECTING

"How do your guys in the field collect stream sediments?" I casually asked the well-dressed man seated next to me at the Alaska mining conference in Fairbanks. I knew his company had been doing reconnaissance sampling of large claim blocks in Interior Alaska near the newly-discovered Pogo gold deposit.

"We collect fines from small eddies or bars in the active channel," Jeffrey Currey, Avalon Development Corporation's senior mining engineer, readily answered.

"In your pan concentrates," I pressed on, "how do you compensate for losing micro-gold?"

"We have a field procedure everyone follows to ensure consistent samples," Currey explained. The speaker tapped the microphone for attention, and began his presentation. "I'll send you a copy," Currey promised. A few days later I received excerpts from the *Goodpaster Project Field Manual* that described the procedures for obtaining stream sediments and pan concentrates.

BENEFITS OF GEOCHEMICAL PROSPECTING

Geochemical prospecting has become one of the most powerful techniques for finding ore deposits. Gold panning, an age-old technique, is essentially a way to examine the geochemistry of heavy minerals in a drainage system. Called pan concentrates, the higher specific gravity minerals give clues to rock types, mineral associations, and metals. Other geochemical techniques analyze stream sediments, soils, rocks, and vegetation. Refinements in the basic techniques are constantly being made to adjust for different sampling environments and to make use of lower detection limits in the search for new deposits.

For pan concentrates, the *Goodpaster Project Field Manual* stressed consistency in the collection site ("moderate energy sites which contain a mix of one-inch gravel to twelve-mesh sand"), the sample volume ("sift down a sample from about one cubic foot"), the size fraction ("ten-mesh or smaller"), the sample weight ("just over a pound"), and the panning method. All of this leads to less sampling error and better data.

Their method for obtaining the final pan concentrate answered my question about saving micro-gold. The negative ten-mesh material was not panned down, thus minimizing the possibility of micro-gold being washed out of the pan. Altogether, four pans of material were sieved from each sample site and then combined for analysis.

Another exploration company is experimenting with keeping a coarser size fraction (say, positive ten-mesh or coarse sand-sized) and grinding it down to release the minerals entrapped in the larger particles. Milt Wiltse, head of the Alaska Division of Geological and Geophysical Surveys in Fairbanks, Alaska, discussed this issue in DGGS *Report of Investigations 88-13*, "The Use of Geochemical Stream-drainage Samples for Detecting Bulk-Minable Gold Deposits in Alaska" (1988).

HISTORY OF GEOCHEMICAL PROSPECTING

Modern exploration geochemistry began in the 1930s when Soviet geologists perfected emission spectrographic analytical methods and sampling procedures for soil surveys. At the same time, Swedish and Finnish geologists began analyzing plants and correlating indicator plants with ore deposits. Geochemical exploration by United States and Canadian companies didn't begin until the 1950s.

In contrast to the Soviet use of the emission spectrograph, the USGS developed colorimetric methods using dithizone and other chemicals to test for color changes from the presence of ore metals. Canadian and Alaskan prospecting programs taught how to use dithizone, and much of my early reconnaissance work used soil sampling with dithizone, a technique I learned in Professor James Madonna's classes at the University of Alaska Fairbanks.

DITHIZONE

"Shake, shake, shake!" he would admonish us as we mixed sieved soil with distilled water and uniodized salt. Then, after carefully dripping a few milligrams of green dithizone (mixed with white gas) into the muddy test tube, we would stopper it and shake some more. If the deep green turned to red, we knew there were minerals. Various shades of red-pink-orange could indicate differences in the metallic elements present. For a cheap reconnaissance technique, dithizone may still be very useful.

"Dithizone is useful for measuring iron in the soils," Professor Madonna conceded when I pressed him about its use. I was frustrated by the limited usefulness and unreliability of the method. Frozen soils would give me a pink color even if no economic minerals were present, and I wondered why the dithizone changed to yellow in another area I prospected. So eventually I switched from dithizone to assays to obtain more precise information.

ASSAY VALUES

I visited Professor Rainer Newberry in his office on the campus of the University of Alaska Fairbanks to discuss some of the issues in geochemical exploration. A large chart of the "Periodic Table of the Desserts" explained that element 70, Au, stood for "pie, crustum Kolokuntes."

"How does a person decide if an assay value is significant?" My question caused a pained expression on his face, and a moment of thought.

"What's background and anomalous varies from district to district and deposit to deposit," he answered. "The standard procedure is to do an orientation survey first, but this presumes you know what you're looking for and you're only looking for one thing. The alternative is to say, 'Whatever shows up, shows up.'"

He ran one hand through his hair, placed elbows on the desk, and attacked the question from a different angle. "From a scientific standpoint, it really is important to work in terms of models and testing models. It's been the approach pushed academically for the past few decades. The problem with that approach is that if we knew everything, we'd know everything."

Professor Newberry taught a class on geochemistry, and here I was asking for a semester of information in a few sound bites. He settled into the task.

"Nothing surpasses the Fairbanks District for a variety of deposit styles," he continued. "People realized, 'Hey, there's a whole bunch of things out here with a whole lot of geochemical signatures.' If you use arsenic and antimony, you're going to miss the Fort Knox Deposit. If you use bismuth, you lose the True North Deposit. Those are the two biggest deposits in the district. It comes back, for better or worse, to having a model."

Returning to my original question, he tied his example in to the problem of background and anomalous values. "Sampling around Fort Knox for arsenic would show a background of ten parts per million, and one hundred parts per million would be anomalous. If you took that same correlation five miles away to the True North Deposit, you'd say every sample was anomalous." He paused and closed his eyes. "The standard approach is to take several samples that aren't mineralized and not weathered, to give you a sense of what background is. But only for that area you've sampled."

A graduate student asked him to examine some thin sections under a microscope. I followed them down the corridor to the petrology laboratory where six microscopes were set up. Newberry studied the thin section, discussing what he saw. The student suggested an explanation of his observations.

"Precisely, precisely!" Newberry agreed.

Back in his office, I asked about pathfinder elements such as arsenic that were often widespread in soils near some types of deposits.

"Part of the whole business about arsenic and gold dates back to the 1960s when it was difficult to get analyses of gold below one part per million," he replied. "Now that they are getting down to five parts per billion, it makes sense to use gold or bismuth rather than arsenic. It tells you what kind of beastie you're dealing with. It's analogous to the situation in platinum group element analysis today. Those assays are high cost, so you'd be looking for associated elements such as nickel, or maybe even copper [in lower cost assay packages]."

ATOMIC ABSORPTION SPECTROSCOPY AND INDUCTIVELY COUPLED PLASMA

The development of atomic absorption spectroscopy (AA) in the late 1950s, and now inductively coupled plasma (ICP), allow rapid, accurate, and sensitive analyses of many elements. Multi-element geochemical assays give a wide spectrum of analyses for over thirty elements, including gold and mercury at the parts per billion level. Yet, Newberry had reservations about blind acceptance of assay results.

"Once you get down to values close to the detection limit, there's a lot of error involved," he noted. "Concentrations for about half of the elements can be worthless because the acids and methods used for dissolving the sample won't be effective for many of the elements. To a certain extent, a company will tell you this in the fine print. The standard ICP geochemical package yields a lot of data, but you have to know what is bogus."

Later, I checked what a major assay laboratory's booklet said about their ICP package. "Only a portion of the more resistive metals is dissolved in this matrix," it stated. "Because of this, these packages are normally used for exploration programs."

EXAMINING ASSAYS

Geochemical exploration works on the concept that pathfinder elements such as arsenic or mercury are mobile and disperse widely in a halo around a deposit, or downhill from an outcrop. Ore bodies that don't outcrop or that are covered by glacial till or soil may have leakages into the water table, or vegetation may concentrate some elements.

"For many porphyry copper deposits there is a zinc halo around them," Newberry pointed out. "But it's at such a considerable distance that it's not tremendously useful."

Ore deposits often have indicator elements, a suite of elements that are often diagnostic of a particular type of deposit. Precious metal veins and Carlin-style disseminated gold deposits would have gold or silver as the indicators, and arsenic, antimony, and mercury as more widely dispersed pathfinders. Porphyry copper indicator elements are copper and molybdenum, with zinc, manganese, and gold as possible pathfinders. Meanwhile, volcanic-associated massive sulfide zinc-copper-lead deposits would have barite and arsenic.

"Platinum group metals (PGM) would have nickel and copper for sure," Newberry noted. "And commonly cobalt."

I asked him if chromium was another good PGM indicator. "Yes, but chromium is one of the elements that doesn't go into solution very well, so standard ICP doesn't dissolve it as thoroughly."

R.W. Boyle compiled these associations from empirical data in "Elemental Associations in Mineral Deposits and Indicator Elements of Interest in Geochemical Prospecting" (Geological Survey of Canada *Paper 74-45*, 1974). Most textbooks have a table of typical ore body element associations.

Newberry concluded our interview by emphasizing fieldwork in spite of the theoretical and laboratory constraints. "It's going out and looking for stuff, even if you are barking up the wrong tree, that counts."

Many texts on geochemical prospecting discuss how elements are distributed in different environments, mineral anomalies, dispersion and mobility of elements, sampling techniques, and methods used in exploration. H.E. Hawkes wrote "Principles of Geochemical Prospecting" for the USGS *Bulletin 1000-F (1957)*, A.A. Levinson did *Introduction to Exploration Geochemistry* (1974), and Arthur W. Rose, Herbert E. Hawkes, and John S. Webb covered similar ground in *Geochemistry in Mineral Exploration* (1979).

DOING FIELDWORK

I had seen geologist Dave Butherus at work collecting rock and soil samples during the worst mosquito and rainy days of summer on one of my prospects, so I interviewed him about his techniques. He began geological work in Alaska in 1974, and spent time in many areas of the state.

"Try to take as many representative samples as you can," he advised me. "Key into alteration, color changes. Is this ground prospective for anything? Get a lot of samples, to see what your background values are. Also, don't be afraid to take some high-grade samples of mineralization to find out which elements or minerals are present. A lot of our sample grids go out on the ridges where it's often easier to get good samples, but this is tricky because recessive areas might have mineralization. In mountainous terrain you can see rubble and talus, so it's easier to tell what geology and mineralization are present."

"A good reconnaissance technique is to follow drainages and beat the heck out of the float, looking for oddball float to follow up. On the first go around, any anomaly is interesting. On a hillside you may have to trench or drill a lot to follow an anomaly up. It's easier with float in a creek." He reflected on his advice. "You're torn because you want the greatest number of samples, but you've got to haul them out. The larger the samples you get, the more representative the results will be."

Behind him were shelves of USGS reports and geology texts. "This initial look at the ground tells you what is background and what is anomalous. Each type of sample has a different meaning quantitatively. If sampling bedrock, you'll normally set higher anomalous thresholds. On transported soil or stream sediment you can be interested in things that are detectable as long as they're above background. If you see any gold in sample results or pans, you'll be interested. In a huge drainage even a small amount of gold is significant because of dilution. Somewhere there has to be a concentration or a large area of low-grade mineralization."

Money limits the quantity of samples you can collect and sample density depends on the size of the target. "In this terrain [Fairbanks and Interior Alaska] we'd do 100 to 200 feet spacing on a soil survey line and run lines 400 feet apart if we were doing a grid. When you get your data back," Butherus noted, "you can use it to tailor your next actions, such as doing closer spacing. You are playing with so many variables that you can't control; in the reconnaissance phase, you're doing broad sampling–a real long shot. Economics are a real stickler. It's not cheap to do this stuff."

I asked him about the usual methods of sampling. "Stream sediments are more comprehensive," he felt. "They are a coarser tool because they represent a much larger area. Soil is the next less specific type. There's a certain amount of dispersion. If you get into bedrock with the soil, you're mainly sampling what's right there. Rock sampling would be the most specific. Pan concentrates, heavies from a stream, don't tell you the concentration of mineralization. They just say it's available. Similar to gold panning, we know there's gold up there, but we don't know the quantity upstream."

He returned to soil sampling. "Up here, we tend to go for the "C" horizon, the zone of weathered bedrock. In the weathered zone there's a certain mobilization of minerals in the soil due to percolation of water. The soil sample is prepared by sieving off and analyzing a certain size–usually the negative eighty-mesh fraction. That's one of the weaknesses. It's hard to get homogeneous samples–depends on how the bedrock decomposes. Some would be big, angular blocks. Others would be a pulverized mass of rocks and soil. This means that the analyzed fraction would be a different percentage of each sample. It takes more effort to get samples out of hillsides to represent local bedrock. Talus and soil move a long, long way. You do the best you can."

SOIL HORIZONS

Soils have long been classified into "horizons," or layers. The humus and surface layer is called the "A" horizon. Colloidal silicates, oxides, and organic material make up the "B" horizon, a zone of accumulation from surface leaching and ground water percolation. "The "C" horizon represents decaying bedrock mixed with soil. Geologists debate which is best to sample–the "B" or "C" horizon. The choice depends on the locality's soil development profile. "In general, the 'C' is what you're interested in up here," Butherus recommended. A mature soil profile doesn't exist. The situation is quite variable. I try to get some bedrock and some soil."

PAN CONCENTRATES

"Pan concentrates are more specific to heavy minerals," he continued. "Lead, precious metals, and tungsten. Stream sediments are a little more general because the fines–clays and soils–are good at adsorbing ions. Very often we look at both streambeds and pan concentrates. It depends on what you're looking for. Base metal surveys often don't take pan concentrates because the minerals show up well in streambeds. Obviously, the more pans the better. I'd take pans from several different spots where I expect to find heavies–below boulders and riffles, a point on a bar. You select from each of these optimal places. There's tremendous variation. Some streams have a lot of fines, others only have cobbles. A grizzly (3/8-inch mesh) saves time somewhere in there. Your concentrations will be fine material.

"Streambeds are similar. You want to get mature sediment, not flood sediment. Dig down into the stream for something that's been riffled. Make a composite. The more places you get it from, the more representative. It's important to describe the sample site—boulders, bedrock riffles, stream size, gradient, and maturity of the stream."

Each environment requires a different method of sampling. "Loess is a really classical problem here," he pointed out. This wind-blown silt from Pleistocene times blankets hillsides and valleys in Interior Alaska. "It's totally inert for all practical purposes. You have to dig or drill through it. Loess is only silica. It dilutes your sample. It doesn't represent bedrock in any way."

He uses a power auger with a three-inch drill to get through three to six feet of loess cover, but if the layer is thicker, heavier drilling is necessary. I carry a pointed shovel for digging shallow pits for most soil samples. Butherus used a narrow, long-bladed shovel for getting samples in western Alaska, and in some places in Interior Alaska.

"Underlying sampling is geology. If you don't know the ground truth, you need to learn the rocks," he recommended. "You can't just sit here and arm wave. You have to go out there and see what you're looking at. The challenge is making sense of it. It rings a bell, and then you go out and push it a little further."

SUMMARY

Tens of thousands of geochemical samples are now processed every year, and many new discoveries are credited to this method. Combined with geophysical and geological mapping, these geochemical exploration techniques locate envelopes and halos around mineral deposits. Created by dispersion of chemical elements from weathering, erosion, gas transport, and other environmental factors, anomalies stand out from background levels.

Regional reconnaissance programs are used by mining exploration companies to identify possible mineralized areas. Other targets to investigate are suggested by stream float, rock alteration, granitic intrusions, vegetation "kill" areas, and gold panning. Follow-up geochemical sampling can narrow down anomalies for more intensive investigation. Modern prospecting has been revolutionized because of geochemical sampling methods and because of lower detection limits in assaying.

16. Rock and Mineral Alteration

Driving past the Lavender Pit just outside of Bisbee, Arizona, I can only imagine Sacramento Hill rising above the gaping hole. Yet, next to the highway opposite the pit is another red-stained hillside that shares similarities with the mined-out copper deposit. It takes a lot of willpower not to pull over and walk along the roadcuts where unoxidized pyrites glitter. Here is a living, rusting example of pyrite breaking down into limonite. German miners called such a reddish area an "iron hat," or gossan. In Bisbee's case, the Red Mountain's copper content is not yet considered sufficient to justify mining.

SIGNS OF MINERAL DEPOSITS

When mineral deposits are formed, the surrounding rocks are often altered by a hot magma, hydrothermal fluids, or gases. Additional changes occur because of near-surface weathering and percolating ground water. Recognizing these forms of alteration helps a prospector discover hidden deposits. Indications of nearby mineralization are resistant breccia pipes forming rusty red thumbs, colorful bleached areas, and depressions where leached sulfide zones have collapsed. On a smaller scale, veins exposed to oxidation and groundwater may lose their minerals, but leave cavities, limonite, or colors that are clues to what minerals were once there.

"Alteration can really change the way a rock looks," explained Sam Dashevsky, a wiry, black-haired geologist with eighteen years of Alaskan exploration experience. "An altered rock may look like a normal rock, if you don't recognize that it's been changed. To be able to recognize altered rocks, you need to have seen enough regular rocks. The first thing to do when you're going into a new area is simply mark on a map all the places where rocks are sticking out. You can do this from air photos, or during an overflight, or as you cover the ground on foot. Examine the rocks and build a basic map of what rocks are where—an outcrop map. Look for signs of alteration and mineralization."

I was visiting Dashevsky in his log cabin office in a grove of birch trees outside of Fairbanks, Alaska. Dashevsky selected a jagged rock with a cemented mixture of sand, quartz, and pebbles from the window sill.

"On my first job for Inco in the Seventy Mile country west of Eagle, we were looking for young, Tertiary-aged hot-springs-related gold deposits. We were exploring in the Tintina Trench, looking at the Tertiary gravels that fill that basin. There was placer gold production from the area and the U.S. Bureau of Mines reported cinnabar, a mineral that forms in the upper levels of an epithermal system."

He turned the rock so the light over the map table illuminated its different sides. To me it looked like a conglomerate, only lighter. It had been slabbed on one side to reveal a pebbly sand saturated with quartz.

"On our first day in the area we were doing an overflight to see where potential outcrops might be. From half a mile away we spotted a gravel bluff with a white and cliff-like appearance. We immediately recognized that for these young gravels to form such a cliff without collapsing, they had to be hardened, silicified.

"We landed a half-mile away and hiked over to it. It was indeed a gravel bluff, but these gravels were cemented by silica and cut by quartz veins. In the following week we found two prospects with significant enough gold mineralization to warrant multi-year drilling and exploration. The heart of the silicified system we were looking at was so altered that on the foggy day when I first stumbled on it, I briefly wondered if it was granite."

I turned the rock over, feeling its sharp edges. A hammer beating on this rock would ring. It reminded me of walking on the sharp sinter, or geyserite opal around the mouth of a still-active hot spring and hearing the shards tinkle underfoot. Besides vein quartz these hot springs systems create jasper, chalcedony, chert, and opal. In a mine, hammering on an altered silicified zone produces high notes, and then the hammer sound deadens when unaltered rock is encountered.

Jasperoid, an important form of silicification, is found in more than two hundred mining districts in the United States. Silica-bearing solutions most readily replace limestone and dolomite. These replacement bodies of jasperoid are usually localized along channelways in faults, shear zones, and breccias, but the solutions carrying silica may also be widely

Widespread rock alteration at the Comstock silver-gold lode illustrates the effects of hydrothermal fluids on the enclosing rocks. Quartz veining and clays characterize the productive center of the system.

Geologist Sam Dashevsky holds an example of silification where circulating silica-rich water cemented a porous conglomerate. Rock alteration is a sign of potential mineralization.

dispersed throughout a porous host rock. Many gold and copper deposits in Nevada and the southwestern United States have associated jasperoid.

"The lithology [rock type] really determines the alteration," insisted John Galey, Kinross Gold Exploration's Arctic geologist with experience in Colorado, Nevada, Greenland, Russia, and Alaska. "In sedimentary areas you get jasperoid. If you're working in andesite volcanics where there's a lot of iron, look at the pyrite content, and the chlorite. It's very rock-dependent. You can't alter a quartzite. Be aware if you are in carbonate terrane. There, you develop a skarn. You hope to see some garnet, some pyroxene, diopside."

If there are calcareous or carbonate rocks such as limestones at an igneous contact, the alteration or "skarn" will create silicates such as garnet, epidote, and wollastonite and possibly copper, tungsten, lead, zinc, iron, or gold deposits. The country rocks around intrusives should be studied for these red- and green-colored contact effects. In the absence of reactive rocks such as limestones, hot magma bakes, and silicifies the surrounding rocks into a tough matrix called hornfels, which is so hard that a hammer just bounces off of it.

"I worked on a molybdenum property in Greenland," Galey reminisced. "I was young and nobody else wanted to go. There, a Tertiary-Age granitic stock intruded shales and siltstones. It had a big brown hornfels zone wrapping around it. Once you walk on hornfelsed ground, you never forget it."

Another well-known form of alteration is the development of white mica or sericite, a hydromuscovite. This is often seen along veins and fractures or permeating altered areas. When water is added to feldspars and other aluminum-rich rocks, the alteration product is sericite in the form of large flakes or a felted layer of fine grains. The chromium-bearing sericite micas, fuschite and mariposite, seen in California's Mother Lode, apparently come from the breakdown of magnetite, pyroxenes, and amphiboles.

"If you've got white mica bordering your veins, you've got [evidence of] hot fluids of an altering nature," Dashevsky pointed out. "Sericitic alteration is a guide for not only gold but any hydrothermal alteration, even volcanogenic massive sulfide (VMS) environments [undersea copper-lead-zinc deposits]. In the waning stage of the development of a VMS system, hot circulating waters alter the enclosing rocks with widespread sericite. An extensive sericite halo may mean you're walking on top of a mineral deposit."

In an ore deposits class at the University of Alaska Fairbanks, we examined multiple veining from the Butte copper system in Montana. The rich "main stage" veins, which formed later and were a few inches to several feet wide and up to a mile or more in length, had an envelope of sericite along their margins. Emplacement of a hot magma heated the groundwater causing convection cells to form and leach minerals out of the magma and surrounding rocks. Sulfides, clays, and aluminum minerals then precipitated out during the cooling process. This sericitic phase of the alteration of a magma destroys the original rock textures and is related to the influx of groundwater. Pervasive late alteration, the quartz-sericite-pyrite mineralogy carries the highest sulfide content in a porphyry copper system.

Sericite with quartz is also found enclosing or above tin, tungsten, beryllium, or molybdenum deposits in fluorine-rich apical portions and contact zones of a granite. Called greisen, this mixture of quartz and sericite (or the lithium mica, lepidolite) is formed during alteration of the granite by hydrothermal solutions.

IGNEOUS ROCKS

Another major type of alteration attacks the feldspars in igneous rocks. This alteration and bleaching of feldspars, and formation of the clay minerals kaolinite and montmorillonite along veins and in the wall rocks is called argillic alteration. Geologists are sometimes called "rock lickers" because of their way of testing these softened feldspars. "When you lick it and your tongue sticks to it, it's argillic," advised John Galey. "If you can't get your tongue off it, it's advanced argillic." Although alteration of feldspars to clay may also be confused with fairly recent surface weathering effects, argillic alteration from hydrothermal fluids is a useful indicator of potential mineralization in igneous rocks.

At the Round Mountain Caldera-related Gold Mine in Nevada, geologists employ dental probes to check the volcanic rhyolite rock matrix for degree of alteration. Areas with soft clay alteration have low-level gold mineralization. Sericite indicates a moderate level of gold, and silicification and hard rocks have a high level of gold. Some geologists take a small amount of the altered feldspar and grind it between their teeth as a field test. The grittiness tells them how much feldspar remains.

"You have to spend time looking at the host rocks in their unaltered state," said Galey. "Look to see if they're bleached white, chloritized, or silicified. Some of it's pretty subtle. It depends on how good you are at recognizing rock alteration products. If you don't know what altered rocks look like, you won't know if it's the original rock texture or a bleached equivalent."

Dashevsky echoed this idea. "When traveling to active or historic mines, it's important to go out beyond the pit and away from the ore zone in order to know what's out there, to know that this could be the tip of the iceberg. Take time to walk to the edges of the mine. Look at the distant signs of the mineralized system. Look at the waste dumps and outcrops. This altered vein could be on the edge of a larger system. Remember, you're only 1,000 feet from a billion dollar deposit, and this is what the rocks look like."

Recognizing the major types of alteration can help a prospector locate the outer edge of a deposit, and aid in the search for the more central areas where argillic, sericitic, and silicified rocks crop out. The best place I've found to look at alteration is the Comstock Lode around Virginia City, Nevada. Rocks near the center of the system crumble into powder, evidence of argillic alteration. I've walked around the pits after a rain shower and had my shoes gummed up from the sticky clay. Pervasive quartz veins that once carried silver sulfides through the silica plumbing of the hot springs system contain amethyst crystals, pyrite, and gray sulfides. In outcrop they cross-cut the country rock in many directions, a sure sign of fracturing and multiple pulses of fluids over time.

Slightly away from the Comstock ore bodies, strongly altered granite is greenish and clayey, and fresh pyrite glitters throughout the rock. This is the zone of propylitic alteration, and it extends up the slopes of Mount Davidson, where fractures in the granitic rocks are tinged green with chlorite and epidote. The Comstock is a classic example of zoning: clays and quartz near the silver veins, and chloritic or propylitic alteration away from the mineralization.

In much the same way as around the southwestern states' copper porphyry deposits, the Comstock system had circulating fluids and heat that penetrated far out into the surrounding rocks. Chlorite and fresh pyrite were fringing indicators of the mineral deposit.

"In the propylitic zone, pyrite is stable," Dashevsky told me "These are classic zones or halos around porphyry copper systems. The alteration envelopes extend well out from the deposits. In the western states, geologists map alteration halos. Some didn't breach the surface. By mapping them systematically and developing alteration patterns, drill targets can be developed."

However, barriers to the fluids can deform the shape of the halo, leading geologists to miss the deposit if they simply target the centers. For example, the Mount Davidson granite acted as a dam along the western side of the Comstock fissure-fault. The silver lodes were found to the east under Tertiary rocks and older volcanic andesite cover.

There are many types of alteration in addition to the major ones mentioned above. Along quartz veins in the granite of the Fort Knox Gold Deposit near Fairbanks, Alaska are white margins of albite, a high temperature alteration known as feldspathization. In other deposits, tourmalinization occurs in veins and wall rocks where boron was introduced. I've seen thin needles of tourmaline, bent from their fluid transport, in schist near mineralization. Masses of black tourmaline accompany some gold-quartz veins.

CARBONIZATION

The introduction of secondary carbonates in host rocks and veins is carbonatization, another kind of alteration. The California Mother Lode veins developed ankerite, a calcium-iron-magnesium carbonate, from a related process known as ankeritization. Calcite is frequently found in many types of gold deposits.

FALSE SIGNALS

Not all alteration guarantees mineralization. I remember going on a University of Alaska Fairbanks Mining Extension field trip to Rainbow Mountain in the Alaska Range to look for fossils. The area got its name from the oxidation of iron pyrite, which colors the gray scree slopes red and orange. As the group hiked up McCallum Creek, we came upon a huge mass of pyrite mushrooming out of the talus, and here and there on the steep slopes other humps of pyrites stuck out.

"Shouldn't there be copper deeper in the system?" we excitedly asked Professor James Madonna, our Mining Extension teacher. We eagerly sank our rock hammers into the disintegrating sulfides looking for minerals.

"These are barren vents from a volcanic system," he explained. He photographed us climbing over the rusty rocks, exploring our first gossan.

The high mountain climate prevented much oxidation, and just under the surface the pyrites were fresh and metallic.

On another UAF field trip, Professor Rainer Newberry led us to one side of Pedro Dome, above rich placer gold creeks and not far from two gold lodes. A colorful zone of yellows, oranges, and reds crossed a road cut and slanted down the mountainside. We attacked the rocks with hammers, searching for minerals in the oxidized material.

"Shear zones such as this one may look mineralized," Newberry warned us, "but they aren't. What you see is surface weathering and not alteration from hydrothermal fluids. The shear zone is porous, allowing surface water to easily percolate through it, turning the crushed matrix to clays and oxidizing the iron. There's no mineralizing event here. It's barren."

In the past I had lugged large chunks of this rock home for study, thinking that with all the gold in the creeks below, this had to be related to the gold source. To add to the confusion, some of the shear zones around Fairbanks do have gold and antimony in them, along with a lot of iron.

"We found a spectacular hydrothermal breccia on Ruth Creek north of Fairbanks," Kinross Gold's exploration geologist Galey recalled. "It was right in the creek where they'd been placer mining. There'd been a hot springs there. We sampled it, slabbed it, and looked at it under a microscope.

Colorful alteration may be from surface water and oxidation, or it may originate from a hot magma and circulating hydrothermal fluids. The Ruth (Ely), Nevada copper porphyry deposit affected a large area in the surrounding rocks.

It had pieces of sepentinite floating in chalcedonic quartz. [And] some pyrite. But we couldn't squeeze any gold out of it."

False signals of mineralization are seen in iron-rich sedimentary rocks, in the red soils in the Sierra Nevada foothills of California, and in pyritic beds formed under reducing conditions. Although iron is a major constituent in most mineral deposits, it also appears in many unmineralized settings. I've seen beautiful pyrite cubes in rocks that were unrelated to ore deposits. Iron and manganese in solution may mix with rock fragments to form a cemented matrix that looks suspiciously like a leached outcrop or gossan.

"Pyrite, in some cases, can lead you astray." Galey spoke from experience. "What else do you see? We get more excited when we see arsenopyrite because in the Fairbanks District it is associated with gold."

SURFACE LEACHING OF MINERALS

After forty years of observations in the southwestern United States and Australia, Roland Blanchard wrote his report, "The Interpretation of Leached Outcrops" (Nevada Bureau of Mines *Bulletin 66*, University of Nevada, Reno, Nevada. 1968). It's an exhaustive look at the processes and products of leached mineral deposits. Blanchard gives several examples of how examination of gossans can yield information about the presence or lack of economic deposits at depth. Oxidation of many minerals creates a distinguishing cellular structure that Blanchard feels can be used as a key to identifying the original mineral.

The empty cubes where pyrite has been leached away are readily recognizable. Often, the striations of the pyrite faces are etched into the rock matrix. Other forms of pyrite are flat limonite in fractures, deep brown or black smeary crusts, or columnar limonite derived from massive pyrite.

Many ores produce distinctive colors during the oxidation process. Hammering on an oxidized vein last summer, I found green, yellow, ochre-orange, and brown colors–pale olive-green from arsenopyrites, canary yellow from stibnite, ochre from galena, and brown from pyrite. I traced the vein's float across a rubble pile of rhyolite by following its colors.

"Here in Fairbanks, you see antimony [stibnite] and arsenopyrite with gold," Galey noted. "You need to know how they oxidize. Arsenopyrites will oxidize to scorodite; stibnite to stibiconite."

As part of their training, geology students at the University of Alaska Fairbanks are taken to an exposed roadcut on Pedro Dome near the turnoff to the Fort Knox Mine. A trail of yellow rocks leads to a lens of stibnite in the schist. The distinctive canary yellow color of the oxidized mineral is hard to miss.

Metallic sulfides that are exposed to air and water are leached away, leaving colors, spongy textures, and cavities. With the right conditions, these minerals are redeposited deeper where air is excluded. At Bisbee, Arizona, during the late stages of mineralization, disseminated pyrite and chalcopyrite were concentrated in the outer shell of the granite porphyry and in the surrounding schists. Subsequently, percolating water created sulfuric acid, which leached the copper from near-surface areas in Sacramento Hill and redeposited it deeper in the system as chalcocite.

In a similar process, the bonanza silver deposits of Leadville, Colorado, the Comstock, Nevada, and Tonopah, Nevada were enriched from minerals that had been precipitated from near-surface ores. The outcrop and upper part of an oxidized deposit may appear barren, while deeper down the secondary ores are enriched.

SUMMARY

"Each deposit will be surrounded by different types of alteration," Dashevsky emphasized. "[In] the early stages of prospecting an area, you can't be locked into any one model. Be open for surprises. Be receptive to luck. Then, you have to follow what you find, wherever it leads you."

Some people believe that all the near-surface gossans and obvious rock alteration areas have been identified. Certainly, red flags such as Bisbee's Sacramento Hill have long since been mined out. However, many of these signposts may still be obscured by ground cover or barely visible until a person carefully prospects the ground. Recognizing the different types, colors, and effects of rock and mineral alteration starts a person on the trail to the discovery of an ore deposit. An "eye for ore" begins with understanding these subtle clues.

17. Yellowstone's Geysers and Hot Springs: Near-surface Mineral Deposits

I joined a crowd that was sauntering through the lodgepole pines toward wisps of steam. Boardwalks and easy trails led us into the Norris Geyser Back Basin in the middle of the famous geyser trail of Yellowstone National Park. Ahead of me a woman wrinkled her nose.

"It smells like rotten eggs!" she exclaimed.

These acidic hot springs and geysers bring large amounts of sulfur to the surface where it's oxidized and forms sulfuric acid. To the prospector in me, that sulfur smell indicates a mineral deposit. Eau de iron pyrites.

Volcanos, and more recently glaciers, created present-day Yellowstone. A huge volcanic explosion 600,000 years ago formed a thirty-mile by forty-mile crater, or caldera, that is now partly filled with Yellowstone Lake. Active geysers and thousands of hot springs are evidence of a near-surface magma chamber of molten rock. The area's past volcanism, rapid uplift, strong heat source just below the surface, and frequent earthquakes emphasize its potential for future volcanic activity. Yellowstone's subsurface hot spot mantle plume is similar to one under Hawaii except that here it has evolved under continental, not oceanic, crust.

I joined other visitors gathered along the wooden benches around Echinus Geyser, waiting hopefully for one of its one hundred twenty-five-foot bursts. We waited and waited, seeing only small clouds erupting.

"It didn't get its vitamins today," someone commented as they were leaving.

I wandered around the beautifully situated pool, noting the spiny encrustations on the rocks that had given the Echinus Geyser its name. Reddish and yellow-brown coloring from iron oxides and arsenic on the sinter and pebbles reminded me of why I had pilgrimaged to this spot. Here was a mineral deposit being formed as I watched its rumblings, bubbling, and squirts. When miners dug into the bonanza gold and silver deposits of the Comstock and Round Mountain, Nevada, they uncovered "fossil" systems that had once shared similarities to Yellowstone's geysers and hot springs.

For economic geologists and prospectors, mineralized hot springs are "living" demonstrations of ore deposit formation in an epithermal, or near-surface environment. These bonanza silver and gold deposits share characteristic breccias, rock alteration types, and trace element compositions. In a similar way, many of the sediment-hosted micro-gold deposits of Nevada's Basin and Range were formed as water that was heated by a deep magma chamber circulated minerals into reactive limestone strata.

John Colter explored the Yellowstone country after leaving the Lewis and Clark Expedition in 1806 and discovered the fantastic hot springs and geysers. Not until 1872 did Congress recognize the unique features of the area and create a park and preserve.

"Yellowstone [National Park] wasn't created because of its charismatic bison, but because of its geysers," Professor Stuart Rojstaczer, a geophysicist and hydrologist at Duke University, asserted during a talk I attended at the University of Alaska Fairbanks. "Geysers are incredibly beautiful and rare. Other places where geysers are found are on Russia's Kamchatka Peninsula, Iceland, and in New Zealand."

The hot springs and geysers of Yellowstone National Park are models of shallow (epithermal) gold-silver deposits. Echinus Geyser in the Norris Geyser Basin produces arsenic and iron.

When I first watched dark bison silhouetted at sunset against the white steam of Midway Basin's geysers, I realized that bison and geysers together are the magic of Yellowstone. A flood of history overwhelmed me as I remembered the role these great animals played in the West. And the fact that seventy-five percent of the world's geysers are found only here also makes this a special place.

ORIGIN AND MINERAL COMPOSITION OF THE SPRINGS

Yellowstone's thermal springs, fumaroles (gas and steam vents), mud pots and geysers demonstrate what happens when water (rain, snow, or groundwater) is heated by a shallow magma. Nevada's Steamboat Springs, south of Reno, and New Zealand's Broadlands are other active geothermal systems that illustrate gold-silver and trace element (arsenic, stibnite, mercury, and thallium) deposition in a near-surface environment.

The Yellowstone caldera and faults provide the open spaces for fluid flow. The caldera has periodic uplift and subsidence related to pressure and leakage from a three- to seven-mile deep magma chamber. Seismic and gravity measurements indicate that the magma may extend at least 200 miles deeper. Surface water percolates downward as much as a half-mile. There it's heated, enriched in chemicals, and driven upward through fractures.

Gorgeous quartz veins with open cavities and ribboning from repeated reopening are evidence of a hot springs environment. Symmetrical banding, comb structures, and crystal-lined openings are all found in these deposits. Yellowstone's hot springs and geysers have clays and widespread alteration in the surrounding rocks, eruption breccias, and abundant silica sinter—additional characteristics of the "fossil" hot spring deposits.

In Yellowstone's geyser basins, geysers, and hot springs discharge silica- and chloride-rich waters at low elevations, creating crystal-clear pools. At higher elevations, mud pots, and nonoverflowing pools have native sulfur because the mud reduces the permeability of the rocks and plugs up the openings. Sulfuric acid is created by bacteria which attack the surrounding rocks, producing yellow, orange, and green clays and muds.

EXPLORING YELLOWSTONE'S GEYSERS

Traveling from south to north, four areas within the park illustrate the variety of Yellowstone's geysers and hot springs. In the southern area, West Thumb Geyser Basin on Yellowstone Lake is a small caldera within the larger Yellowstone caldera. Seventeen miles west, but also within the Yellowstone caldera, the Upper and Lower Geyser Basins (including Old Faithful) contain two-thirds of all known naturally erupting geysers. Located outside the rim of the Yellowstone caldera thirty miles north of Old Faithful, Norris Geyser Basin has the highest reservoir temperatures. Glacier action has revealed its deeper chalcedonic roots. And, on the northwest edge of the park, Mammoth Hot Springs has nearly one hundred hot springs forming limey travertine deposits in terraces.

I began my exploration of Yellowstone's hot springs deposits at the West Thumb Geyser Basin on a circular bay west of Yellowstone Lake. Here, the lake flooded the southeastern part of the ancient caldera, and twelve feet below the lake floor temperatures are just above boiling. Fortunately, a thick layer of impermeable rhyolite inhibits heat exchange between the magma and the surface, or the lake would be a witch's cauldron. High pressures within the plumbing system and the self-sealing by silica being deposited prevent the lake waters from mixing with the shallow hydrothermal system.

Known more for its hot pools of clear, alkaline water, West Thumb's thermal activity takes place in a small area between the highway and lake. Fishing Cone Hot Spring and several small silica cones are within the lake, and Lakeshore Geyser is submerged. A decline in water levels in the 1970s has seen a corresponding drop in geysers, and even the mud pots are less active.

Leaving the placid West Thumb area, I drove west on the forested park road to the main attraction, Old Faithful, and an amazing fairyland of geysers in the Lower and Upper Geyser Basins. Still within the ring fracture zone of the Yellowstone caldera, these geysers are springs fed by boiling waters from deeper, high-temperature reservoirs. Their eruptions can be influenced by earthquakes, atmospheric pressures, gravitational attraction from the moon, and slight squeezing and stretching in the earth's crust.

OLD FAITHFUL

Like so many visitors to Yellowstone, I immediately headed for the visitor center near Old Faithful to get a fix on the next eruption. A blackboard noted that Old Faithful had an interval between forty-five to one hundred ten minutes, with eruptions being one-half to five minutes long. Within ten minutes of 4:40 p.m. was the next estimated activity.

"Remember, we just predict, we don't schedule," the board reminded visitors.

I asked park naturalist, Mark Winner, how they figured the eruptions times for the many geysers in the area.

"One of us goes down into the Basin at 9:00 am to see how they're looking," he told me. "We call in whether they're full or empty, and if the runoff is light or heavy. Grand Geyser, for example, takes a few hours until it fills. We look for overflow. If it just went off, then it takes time to refill."

The young, lean naturalist began to warm to his subject.

"Now with Daisy Geyser, if you get there within ten minutes of when it goes off, you can tell it just went off because of water runoff. It also bubbles through vents ahead of time. That's another indicator."

After fielding questions from visitors about road conditions, where to see moose (the Grand Tetons are better), and elk herd locations, he returned to geysers.

"This area has the most extensive system of boardwalks," he enthused. "The boardwalks are like Disneyland. They take you through fantastic scenery. Daisy would be a two-mile walk from here. If I were going to spend time," he confided, "Doublet Pool is where I'd go. It thumps and reverberates from underwater bubbles. It's deep, with a rich blue color."

This wasn't the first time I would find an appreciation of Yellowstone's geysers among the staff and employees of the park. In the evenings, I met concession employees out walking the trails or sitting by a favorite geyser or pool. People who are serious call themselves "geyser gazers."

Bearded Snuffy-Sam-look-alike park naturalist John Rhoades joined in with his vote for Plume Geyser.

"Plume Geyser goes off every forty minutes," he noted. "Most people don't have a lot of time, so that's where I take them on my walks. It's a beautiful sight."

When I said I was interested in the mineral deposition around geysers and hot springs, Rhoades launched into a calculation they'd recently done.

"Someone said sixty-five pounds of siliceous sinter comes up with the eight thousand gallons of water Old Faithful spews out every day. [But] we calculated the geyser would contain only nineteen pounds of sinter, at a rate of three hundred milligrams per liter. Assuming it's all deposited, [the buildup] probably falls somewhere in between the two estimates," he concluded.

HOW GEYSERS WORK

Two possible methods for generating geysers have been advocated. The textbook model has water from deep within the system flowing upward and filling a shallow reservoir where the surrounding rocks are already hot. A rapid change of some of the liquid water to steam occurs then shoots the water and steam upward. This "constricted pipe" model has a shallow reservoir or reservoirs being discharged and refilled, and "throttle" and "trigger" mechanisms leading to periodic geysering.

The "Old Stuart" model, presented by Professor Stuart Rojstaczer, doesn't see the need for a shallow reservoir. A constricting zone of intensely fractured rock converts some of the liquid to steam as it moves toward the lower pressures of the surface. With a narrow channel, steam becomes the only way to transport the heat.

"Geysers are so scarce," Dr. Rojstaczer felt, "because the [necessary] physical conditions in nature are rare. In Yellowstone these features come together. Most other areas only have fumaroles or hot springs.

"Geyser behavior can be regular, bimodal [two different pulses, one high and the other lower, as Old Faithful exhibits], or with small amounts of chaos. An interval between eruptions is a typical characteristic of geysers."

Soon after I watched Old Faithful's geyser plume reach 120 feet and subside, I got into a conversation with a park concession employee. James Evrard frequently whipped out a dog-eared notebook as we talked and noted geyser eruptions up on Geyser Hill.

"Every year I come out here and make a new formula for predicting Old Faithful," he explained. "Its average interval has lengthened. Now it's mainly only one long interval, not short ones. One reason Old Faithful is so predictable is its isolation from the other geysers in this basin. People don't realize that this whole area is a geyser basin, it's not just Old Faithful."

Behind Old Faithful is Geyser Hill, and farther along the Firehole River are magnificent pools and geysers with names describing their main features: Castle Geyser, Grotto Geyser, Beehive, Lion (roars), Aurum (golden color), Grand, Plume, and Giant to name only a few.

"With geysers, there are no absolutes," Evrard explained. "In 1997, Giant threw out glassy, green rocks from its plumbing system. Norris Pool expanded last summer, and went from muddy to a clear blue."

Geysers in the ten-mile stretch from Upper (Old Faithful, Geyser Hill, and others in a square mile area near the majestic Old Faithful Inn) to Midway and Lower Geyser Basins have a variety of eruption durations from twenty to forty-five seconds at Anemone (three to six feet high every seven to thirteen minutes) to hours and days at the almost-continuously erupting Clepsydra (twenty to thirty feet high).

Dr. Rojstaczer has noted changes in California in geyser eruptions before and after earthquakes. Supporting his observation is Sapphire Geyser in the Upper Geyser Basin of Yellowstone. It was originally a small boiling spring, but after a 1959 earthquake, Sapphire erupted explosively, enlarging its pool and scattering blocks of sinter around the area. Dantes Inferno, a silica-rich hot springs south of the Norris Geyser Basin, began after the same 1959 earthquake.

MINERALIZATION FROM HOT SPRINGS

A few miles north of Old Faithful, the Fountain Paint Pot walk in the Midway Geyser Basin has hot springs, geysers, fumaroles, and mud pots. Episodic geysers such as Fountain, Morning, Clepsydra, Jet, Spasm, and Jelly show a variety of eruption patterns from their plumbing systems. Midway's clear hot springs and geysers contain chloride-rich water flowing unrestricted from a reservoir.

North of Midway Geyser Basin and located just outside of the Yellowstone caldera, Norris Geyser Basin has the highest reservoir temperatures. The Loop Trail and the Back Basin Trail from the Visitors' Center illustrates some of the different chemical compositions of these thermal springs and geysers.

On the Loop Trail to Porcelain Terrace, chlorine-rich water flows directly to the surface from its source, covering the area with a brilliant white coating. A 1,000-foot deep drill hole at the south side of the terrace found temperatures of 460.4° F or 238° C. Cyanobacteria color some areas of the terrace a bright green color. The neutral pH waters allow yellow, red, and brown organisms to grow.

In contrast to the white Porcelain Terrace, Norris Back Basin waters mix with native sulfur at shallow depths and are acidic. On the wooded trail to Back Basin, I paused at Steamboat Geyser in hopes of a repeat of its eruption to 360 feet in 1991. Extensive red iron oxides coated the surrounding rocks, but only a few weak plumes of steam appeared, so I moved on.

Nearby Echinus Geyser, my main interest, was supposed to erupt every hour, reaching thirty to ninety feet. Fragments of banded marcasite and pyrite are sometimes ejected during larger eruptions of Echinus. However, it was May, and groundwater from recent snow melt was probably diluting its hot waters, reducing its explosive power. Analysis of the iron sulfides has shown that they are enriched in arsenic and stibnite. Another couple of thousand years, and I could stake a mining claim here.

Out on the Norris Basin's barren zone, I cautiously looked over Porkchop Geyser Crater. A hydrothermal explosion here in 1989 illustrated how breccias can form around hot springs. Porkchop changed from an intermittent spring to a geyser, and then to a spouting geyser before it

Mineral specimens in the Keck Mineral Museum at the University of Nevada at Reno show the bonanza ores of Nevada. These near-surface deposits often contain quartz veins with open cavities and ribboning from repeated hydrothermal activity.

exploded and scattered sinter debris. Similar explosion breccia fill near-surface necks and channels in epithermal ore deposits such as the Round Mountain Mine in Nevada. Such sinter breccias serve as indicators of the bonanza ore deposits hidden deeper within the system. Pathfinder elements such as arsenic, cinnabar, and stibnite found as coatings on the sinter and within the silica gel are also clues to deeper mineralization.

Just inside the northern entrance to Yellowstone National Park are about a hundred hot springs forming colorful travertine terraces at Mammoth Hot Springs, the world's largest carbonate-depositing spring system. At Main Terrace, the large amounts of water and colorful algae in the springs and runoff channels create the active hot springs called Blue, Canary, Jupiter, Main, Naiad, Trail, and New Blue.

It's an easy walk past mushroom-like Liberty Cap, a dormant hot spring cone almost forty feet high, up along the rock cascade of Palette Spring to Minerva Terrace's ornate formations. Thermal water at Mammoth is heated by magma in a fault zone. As the water moves through the fault it is enriched in calcium and bicarbonate from the enclosing limestone rocks. When the water reaches the surface, the confining pressure decreases like a cap being unscrewed from a bottle of soda. Carbon dioxide escapes as gas and the bicarbonate combines with calcium to precipitate as travertine, a calcium carbonate.

Travertine deposits form in a variety of shapes. Terracettes or semicircular ledges are created around slowly rising pools. Cone-shaped mounds such as Liberty Cap and Devil's Thumb emerge from a single spring and create a massive statue before ceasing to flow. Linear mounds of travertine from numerous hot spring vents along a fracture zone build up fissure ridges. And adding further variety, algae and bacteria in the travertine deposits color these formations in pastel hues.

I spoke with park ranger Bill Wise at the old Fort Yellowstone/ Mammoth Visitor Center desk. He had been working around Yellowstone all winter, and he characterized the different centers of activity as "little villages." Each area of Yellowstone has walking tours, talks, and a visitor center.

"You have to get out and walk around to really see the park," he advised me. "Here, there are evening walks around old Fort Yellowstone, programs at the amphitheater, and guided tours of the terraces."

SUMMARY

During my trip I saw colorful examples of mineral deposition, photographed picturesque elk and bison, and went geyser gazing. Yellowstone country is a magnificent geologic work in progress that illustrates a variety of hydrothermal systems created by a shallow magma. The geological processes at work here are the key to understanding precious metal deposits formed by hot springs.

18. Skarn: Where the Magma Hits the Marble

SKARN MINERALOGY

As I was entering the University of Alaska at Fairbanks science building I saw my geology professor unloading flat cardboard boxes from his car. When I offered to help, he loaded me up with four of the boxes. The lab tables were already overflowing with more stacks of the stuff when I got there.

"No, not mega pizza for everyone," Professor Rainer Newberry announced. "This is drill core from a tungsten skarn near Bishop, California." He looked in several boxes before selecting one for discussion. "Gather around," he directed us. Holding up a thin rock cylinder of drill core, he surveyed the assembled would-be geologists. "Garth, what do you see in this?" We all relaxed while tall Garth unwound himself from his seat and examined the dusty core.

"Rusty, heavy. Greens and reds." An agonized pause. "Garnets, pyroxene?"

"Pre—cisely!" Newberry beamed. "And look at all the massive pyrites. Typical skarn mineralogy. Your lab assignment is to choose a drill hole and log the core. Each box has five rows, representing a total of ten feet." A quick count of the boxes showed fifty to one hundred feet of core from each hole. "Your job is to sketch the changes from the granite contact through the garnet-pyroxene zone to the unaltered marble front. Estimate the tungsten grade, mark the faults, and note any other sulfides present."

He raced to the chalkboard and began his lecture on skarns. Newberry was Mr. Skarn himself, having published numerous papers on the subject and conducted field trips to skarns most of his teaching life. "When high temperature magmatic fluids come into contact with carbonate rocks, all kinds of things happen," he began. "Diagnostic minerals are massive amounts of garnet and pyroxene—reds and greens. Other names for skarns are contact metamorphic and pyrometasomatic deposits and tactites." He paused as if letting us in on a secret. "Skarns can show up two miles from the parent pluton. And they come in an amazing variety: copper, tin, moly, iron, tungsten, lead-zinc, and gold."

I had visited the Mount Hamilton Gold Skarn near Austin, Nevada, before it closed because of the low grade of the gold. Located high on the steep face of a range front, the deposit had been created at the edge of a large granite stock. Rock hammers bounced off the hardened and hornfelsed skarn. Sharp rocks tore up the heavy equipment tires and caused maintenance problems. Much of the deposit was in wollastonite, a white, granular mineral formed by the alteration of marble into a calc-silicate. Boulders of light green pyroxene added some color to the open cut, but with gold concentrations in the one part per million range, shiny sulfides were the only visible mineralization.

Newberry had suites of rocks from the different types of skarns for us to place in their order of formation and distance from the magma. A high temperature mineral such as molybdenite is commonly located closer to the igneous rock, while other minerals are deposited at lower temperatures farther away. Scheelite, a calcium tungstate, is precipitated by reaction of tungsten-bearing fluid with marble or limestone, and it tends to be concentrated at the marble front or disseminated in the garnet or pyroxene zones.

For two weeks we sketched our observations of mineralization in the small cylindrical cores on sheets of paper, noting changes from massive garnet to pyroxene to wollastonite to marble. Graduate students circulated to offer help. Garth was kidded mercilessly about missing the pyrites in the core.

"Hey, Garth, what's this heavy stuff that's kinda rusty?"

The graduate students shook their heads at the monotonous core. "Use your hydrochloric acid," they'd say when asked about a questionable area. "If it fizzes, it's the marble front. If it doesn't, it's been altered to wollastonite." We all spent time with boxes of core in a dark room, passing the ultraviolet light over them, noting white flecks of scheelite, and estimating the grade of the deposit. The analysis had been difficult, and no one was sure they'd done it right. Skarns aren't fun in the field; they are erratic and have to be carefully mapped. Logging drill core is one of those tedious tasks geologists do to bring order to the subsurface geology.

VISITING TUNGSTEN SKARNS

The first skarn I examined was at the Spruce Hen Tungsten Claim near Fairbanks. As an exercise, my UAF Mining Extension teacher, Jim Madonna, had us map the open cut. At one end of the shallow open pit, a wooden headframe squatted over a dark hole in the ground. During the two world wars and the Korean War, tungsten prospects had enjoyed a brief importance. Now, no one seemed to care much about the mine, and it was abandoned. We mapped a narrow zone of garnet, pyroxene and dark green epidote crystals that hosted the tungsten. A thin bed of limestone had been totally replaced by these colorful calc-silicates. However, under a ultraviolet lamp the low-grade specimens showed only a few flecks of scheelite.

I had always wanted to visit the Bishop tungsten mines located in the Owens Valley of California. When I approached Professor Newberry about my trip there, he pulled out the California Division of Mines *Special Reports 47*, "Economic Geology of the Bishop Tungsten District" (1956), and *Special Reports 48*, "Economic Geology of the Casa Diablo Mountain Quadrangle" (1956), and pointed out his favorite places. Northwest of Bishop were the Tungsten Hills with glory holes at Deep Canyon and at the Round Valley Mine. West of Round Valley, the famous Pine Creek Mine was located in the steep Sierra Nevada Mountains, and to the northeast off U.S. Highway 6 near Benton Hot Springs, he told me, the Black Rock Mine presented a magnificent view of its contact zone spread across a mountainside.

"I wish I could go with you," he admitted. He waved his hands in the air. "You'll see these huge roof pendants of metamorphic rocks in the Sierra batholith. Mega contact zones."

"And big piles of specimens?" I asked.

"Pre—cisely!"

I drove south from Reno along scenic U.S. Highway 395. On one side the deeply furrowed east slope of the Sierra Nevada Mountains swept upward, while the pungent sagebrush hills of the Basin and Range spread out to the east. I had often backpacked up into the Sierra from this side, puffing up endless switchback trails to reach granite-bound lakes with their resident cutthroat trout. The bare ribs of the granite batholith that forms the core of the Sierra Nevada are exposed in the high, rounded mountains enclosing the Donner, Tahoe, Carson, and Tioga Pass highways. In places, metamorphic rocks were isolated and surrounded by the granitic magma as it worked its way upwards. Uplifting and erosion has removed most of the overlying beds to reveal miles of light-colored granite with only a few remnants of the older metamorphic rocks.

PINE CREEK MINE

Near Bishop is the Pine Creek Pendant, a metamorphic mass of marble, schist, and quartzite that hosts the tungsten mines of the area. Also near Bishop is the Round Valley septum, a smaller island of metamorphic rocks in the Tungsten Hills where additional mines are located. Marble, called "lime" by miners and prospectors because it consists of recrystallized limestone, makes up the most important ingredient in the formation of tungsten deposits. The California Division of Mines *Special Reports 47* by Paul Bateman on the Bishop area states, "The universal requisite for the presence of a tungsten skarn is the juxtaposition of calcareous and granite rock–a clean calcareous marble and a light-colored granite" (Bateman, p. 19).

Beds of white to blue-gray marble as much as 3,000 feet thick are found in the "islands" of metamorphic rocks in this area. Contact with the granitic melt created skarn, an old Swedish term for the associated silicate minerals, which are: light-brown to reddish garnet, grayish-green pyroxene, olive-green epidote, and quartz. In addition to the silicates, minerals such as fluorite, idocrase, amphibole, calcite, wollastonite, zoisite, and feldspars were formed. White to yellowish or pale olive scheelite was disseminated within the skarn or in rich ore shoots. Also, other metamorphic beds of schist, quartzite, and impure marble had been transformed into a hard, dense rock known as hornfels when they came in contact with the hot magma.

The tungsten mines had their greatest production during the First and Second World Wars and were brought to life again during the Korean War. The Pine Creek Mine, which was operated by Union Carbide's U.S. Vanadium Company, then by Strategic Minerals' U.S. Tungsten, and more recently by Avocet Tungsten (see *Rock and Gem* Magazine, December, 1997: "The Upside-Down Mine" by Joseph Kurtak), is the largest deposit in the region and the most productive in the United States. It's located eighteen miles northwest of Bishop, California at the end of the Pine Creek Road.

By 1953 the Pine Creek Mine had produced $88,000,000 of tungsten, along with significant molybdenum and copper. The ore bodies outcrop at an altitude of 11,200 to 11,900 feet and the deepest workings are at an altitude of 8,130 feet. The mill at the head of Pine Creek used flotation separation, and gravity and chemical processes to obtain scheelite, molybdenum, and copper. The scheelite occurs as finely disseminated grains, which fluoresce white to yellow, indicating some powellite, a calcium molybdate, in the scheelite. The main ore body averages 500 feet long and 75 feet wide over a vertical distance of almost 3,100 feet.

I stopped at the Millpond Campground just north of Bishop and met Bill Curtis and his wife, who host the campground. Dressed in faded blue jeans and a shirt with pearl snaps, Bill sat in the shade in front of their trailer and told me about life at the mine.

"I came into the valley in 1949. It was twenty-four hours a day, seven days a week then. I had numerous jobs at the mill–batch mix, digesters, and presses. Worked on the bull gang–that's the laborers. I mainly drove trucks hauling dirt, and worked on the roads going up to the mine." "They had a couple of snow slides that took out their garage and some of the mill. It's a very avalanche-prone area. Zero-Level was where I was going in when I began work. A-Level was right on top of the mountain. There's miles and miles of tunnel in that mountain. I've heard guys say it took them two hours to get where they were working." He shook his head at the memories. "The early Shirley Temple Mine is up there. It burned down. It was one of the things her dad spent her money on. It never really got off the ground."

I drove up Pine Creek Canyon the next morning just as the sun began illuminating the Wheeler Crest. Wheeler Crest is a sheer face of Sierra granite named for Lieutenant George Wheeler, head of the United States Topographical Survey of this area in the 1870s and 1880s. Aspen along the creek were turning golden, and a few hikers were at the trailhead below the mine adjusting their packs for the long series of switchbacks ahead.

Avocet Tungsten's chief geologist at the mine, Brian Norris, took me through florescent-lit offices to a meeting room with wooden captain chairs. "I've worked in skarns for twenty-one years," he told me. After a short search in a dark closet, he presented me with a small, but very heavy rock. "It's loaded with tungsten. Typical andradite garnet, pyroxene, and calc-silicates." He took me through a few empty offices. Timesheet holders had the names of past employees: Ken, Rudy, Hans, and Ray. "We're down to nineteen on staff," Norris noted. "The company laid off over thirty people last week."

The Pine Creek Tungsten Mine near Bishop, California is located on the eastern edge of the Sierra batholith. When the magma contacted carbonate strata in the overlying metamorphic rocks, a tungsten skarn developed.

Chinese tungsten production has flooded the market. The Pine Creek Mill processes material from other countries including China, rather than mine deep in the mountain here. Also, Avocet Ventures, a London-based company, shipped in concentrates from tungsten mines in Peru and Portugal. Inactive since 1991, the Pine Creek Mine has large reserves that could be developed if the demand for tungsten increases.

Joseph Kurtak wrote *Mine in the Sky*, a history of the Pine Creek Mine that includes personal recollections of some of its miners (order from him at 5621 Whispering Spruce Drive, Anchorage, Alaska 99516. The price is $20 plus $3 for postage). The many photos and personal accounts in the book show his family's long experience with the mine. His father, Ray Kurtak, was mill superintendent, and he developed processes that are still used today.

TUNGSTEN HILLS

Eight miles west of Bishop are the Tungsten Hills, sited at a lower elevation and more accessible than the Pine Creek deposits. Gold placers on the south side of the hills in Deep Canyon contained a heavy white metal in the concentrates that was identified as scheelite. After an eighteen-month search, a prospector found the scheelite in an outcrop of garnet where he'd shot a jackrabbit—hence, the Jackrabbit Mine, which dates to 1913. The townsite of Tungsten City was laid out nearby, but was never developed.

Tungsten Hill mines are grouped around the Deep Canyon (Tungsten City) area and the Round Valley Mine, the oldest and most productive in the hills. Around Deep Canyon are the Jackrabbit, Little Sister, Aeroplane (the first to produce ore in 1916), and Tungsten Blue (Shamrock) Mines. On the north side of the hills, the Round Valley Mine began production the next year. Near the Round Valley Mine is the Western Tungsten Mine and Tungsten Hill Mine. These all had open pits as well as some underground workings.

Both sides of the Tungsten Hills can be reached by using the Ed Powers Road, four miles north of Bishop. Shortly after turning west off U.S. Highway 395, I turned right on South Round Valley Road, passed the Millpond County Park and Campground, and about four miles after the turnoff, located the Owens Valley Conservation Camp. A dirt road led around behind the camp, and I stopped near a back gate to the camp rather than continue on through a sandy-looking wash. A one-mile hike brought me to the Round Mountain and Western Tungsten Mines' green and red open pits on the hillside. A metal building at the foot of the hill contained drill core from an extensive drilling program by Inspiration Mining, and their company drill roads zigzagged across the hillside. There were plenty of light-colored skarn rocks on the dumps and in the pit walls.

To visit the east side of the Tungsten Hills, I followed Ed Powers Road, driving one mile west from U.S. Highway 395 to Tungsten City Road. The wide dirt road went two and one-half miles into Deep Canyon before climbing up to the mines. I parked when the road began to climb because four-wheel drive is needed to get up the steep mountain roads. Hiking up to the Shamrock Mine, I found skarn loaded with iron pyrites littering the track. This skarn and the open pit with tunnels leading into the mountain were heavily iron-stained. Many of the hills around this area have old glory holes with colorful specimens of skarn.

BLACK ROCK MINE

"Don't miss the Black Rock Mine," Professor Newberry had told me. So, to complete my trip to the area, I drove sixteen miles north of Bishop on U.S. Highway 6 to Benton Station, then two miles west on U.S. Highway 120 to Benton Hot Springs. I stopped at the tree-shaded store and asked directions to the mine. I also made the best decision of the day when I asked about having a soak later. "I'll fix up a tub for you," Bob, the owner, assured me.

From the hot springs I took Yellowjacket Road south, and within a mile, it turned to well-graded gravel. After eight miles of gravel, it joined an old paved road (unmarked, but named the Black Rock Mine Road) that went one mile west to the mill site in the shadow of the mine. In 1917 scheelite was discovered here, and by 1928 production began. During 1953 it was the sixth largest tungsten producer in the United States. Now there are four large open cuts near the top of the mountain and more than a mile of underground workings.

After parking near the abandoned mill, I gathered a hammer, sample bags, water, and a picnic dinner in my packsack and began to hike. Flooding had washed out the lower part of the road to the mine, so I crossed the gully on foot and started up the abandoned road. Hiking up a gentle grade, I passed by tunnels and small open cuts, but my goal was the glory hole near the top of the mountain. I heard Professor Newberry's voice whenever I wavered in my resolve: "Black Rock is the best example of skarn. You've got to see it!"

That evening, huge boulders glittering with garnets, deep green pyroxene, and epidote crystals served as my dinner table. I was surrounded by the brightest and deepest colored silicates I'd ever seen in a skarn. I filled my pack with some of the heavy beauties, and descended the mountain as alpenglow lit up the mountains with golden light.

I thought about the striking array of colors and minerals I'd seen. Scheelite, silver-gray molybdenum, brassy chalcopyrite, and pyrite set in a rainbow of garnet, pyroxene, and epidote—classic skarn specimens. The dry Owens Valley and mountains around Bishop, California have many accessible places to collect and examine the unique geology of tungsten deposits.

19. MINERAL FASHIONS

DEMANDS AND TRENDS

World metals markets, technological breakthroughs, and supply and demand influence which mineral deposits are in favor and can attract the financing needed to make a mine. A prospector has to consider the future needs of industry and society and focus on the high-demand minerals. Anthony Evans pointed out in *Ore Geology and Industrial Minerals* (Blackwell Scientific, 1996) that industrial minerals are by far the largest market today, but there is "never talc fever or [a] sulphur stampede" (Evans, p. 5). Being able to recognize a variety of mineral deposits and their potential markets greatly increases a prospector's chances for success.

Unfortunately, predicting trends in the price of minerals can only be done over short periods because political, economic, and environmental conditions constantly change. For example, the price of tin dropped when the International Tin Council Cartel was no longer able to regulate producers. Tin prices had been controlled by the International Tin Council operating through the London Metal Exchange until Brazil flooded the market in the 1980s and the price dropped, forcing mine closures in Bolivia and Malaysia. Without an increase in tin consumption, the continuing Brazilian production lowered the tin price further and made cassiterite, the mineral of tin, a poor target for prospecting.

Tungsten reached a peak of production in the United States in the 1950s, but since then availability of China's vast tungsten supplies has depressed the market, and many tungsten mines in the United States have closed. Uranium prices were high in the 1950s and 1960s, but then dropped because of environmental concerns and a shift from nuclear power back to coal. Now enormous new Canadian and Australian uranium deposits dominate the market.

During and after World War II the demand for metals was high, but by the mid-1970s the increased use of plastic and ceramic substitutes led to reduced metal markets. Water pipes were made of plastic instead of lead. Automobile gas was lead-free, and many engine components were ceramic-based. In addition, recycling of aluminum, copper, zinc, antimony, and platinum group metals increased, and synthetic diamonds, rubies, and sapphires replaced a portion of the market for natural gemstones.

"In the late 1960s everyone was looking for large, low-grade [deposits of] copper," Dick Swainbank, State of Alaska Mineral Development Specialist, observed. "Gold was pegged at $35 an ounce, and silver at about $1," he said. Just out of school in 1970, Swainbank had worked for Resource Associates of Alaska staking claims and doing airborne geophysics.

"In the early 1970s the price of uranium went into the $35 to $40 range. Geologists dropped what they were doing and looked for uranium. The late summer of 1975, I was out in the Mount Prindle country doing a helicopter scintillometer survey. We identified the target within a few hours. That fall I was busy staking one thousand eight hundred claims for uranium. Then, in 1979 Three Mile Island happened and the price of uranium went into the dumpster."

I had first met Swainbank when he gave a presentation to my geophysical exploration class, and he spread out magnetic survey maps for us to look over. Rightfully called the father of geophysical mapping in Alaska, he had long advocated doing airborne surveys to unlock the hidden subsurface geology. At mining conferences and from his office, he has worked to educate people about Alaska's mineral opportunities.

"In the early 1980s there was a short period where people were looking for tin," he explained and pulled out a copy of *The Northern Miner*, a weekly Canadian mining newspaper. "Tin is $2 to $2.50 a pound now. My view of tin is that it will increase in price as the Far East goes down in production. Then the smaller deposits on the Seward Peninsula and Kougarok drainage might be viable."

Swainbank reeled off other metal prices: "Lead was $.21 a pound, zinc was $.49, copper was not even $.81. It was only $.71 last year for the average. I wonder what moly is?... It was $2.50 to $2.70 a pound for oxide—that's not too great. Antimony was $1,200 a ton. That's $.60 a pound. Bismuth was $3.90 a pound. Uranium is $7.50 a pound—a long way from the heydays."

HELPFUL REFERENCES

Our daily newspaper devotes a small section to the prices for precious metals (gold, silver, and platinum) and non-ferrous metals (copper, lead, zinc, tin, and aluminum), but for a more complete list I use *The Northern Miner* or monthly trade magazines such as the *Engineering and Mining Journal*, *Industrial Minerals*, and the *California Mining Journal*.

Every year the State of Alaska Division of Geological and Geophysical Surveys (DGGS) publishes a report of the previous year's mining exploration, development, and production activity in Alaska. For more information, write to:

DIVISION OF GEOLOGICAL AND GEOPHYSICAL SURVEYS
794 University Avenue, Suite 200
Fairbanks, Alaska 99709-3645
(907) 451-5000 / (907) 451-5050 Fax
http://www.dggs.dnr.state.ak.us/

This free publication gives a good summary of where exploration money is being spent, and currently has descriptions of more than one hundred thirty mines and deposits, including the world-class Red Dog (zinc), Greens Creek (zinc, lead, silver, copper, gold), and Fort Knox and Pogo (gold) Mines.

"I call it the report from Hell," Swainbank, one of its authors, remarked. "I've been part of it since 1989. We send out over one thousand questionnaires. The total value of Alaska's mineral industry during the 1998 calendar year, including exploration and development investment and mineral production, was about $1 billion. Roughly sixty percent is the Red Dog Mine. An unbelievable deposit, and they just found a new deposit six miles away–fifty feet of thirty percent zinc!"

An oak display case in Swainbank's office had examples of mineralized rocks from around Alaska. Massive chunks of gray-brown sphalerite from the Red Dog Mine, shiny metallic blades of antimony, and red clots of cinnabar caught my eye. The walls were covered with land status and geophysical maps. One map showed interlocking claim blocks around the Pogo gold deposit. Nearby, a graph of gold prices was spread out on a drafting table.

ALASKA'S MARKET PROSPECTS

"I remember palladium at $5 an ounce," Swainbank remarked. "Platinum was always up there with the price of gold, and because the price of platinum was so high, they went to palladium for catalytic converters. But the supplies from Russia are insecure, and now the president is pushing for more catalytic converters. The demand side is out of sight, so Alaska is a prospect for platinum group metals. There's a lot of interest now. We've had a dozen companies coming by for information."

Of course, today's hot mineral may suddenly be replaced by a cheaper substitute. New occurrences or larger deposits may be discovered. Lower concentrations may be minable as the metallurgy for extracting the ores improves. For example, demand for silver exceeded production in the 1950s, but by the 1970s several factors had improved the supply of silver: the discovery of Kidd Creek in the 1960s at Timmins, Ontario; increased supply from Nevada's mines; the U.S. Bureau of Mines development of a process for recovering silver from film solutions in 1965; the end of U.S. silver certificate redemption, and the government guaranteed price of $.90 per ounce in 1972. Recycling increased supply, and political decisions led to silver losing its favored position as a primary target for prospectors.

The other author of the annual *Alaska's Mineral Industry*, Dave Szumigala, met me at his Division of Geological and Geophysical Survey office. A low cabinet near the door was covered with chunks, slabs, and pieces of split core from gold-bearing granites in interior and western Alaska.

"Look at the specks in this," he said and handed me a piece of Fort Knox granite. At first I only saw gray metallic spots of bismuth, but then I turned the slab in the light and golden flecks appeared. "I was out to the mine on a tour in late January," he explained. "It was -40°F when I pulled this out of a snowbank in the pit."

Every year when I receive a copy of the mineral report, I check the different areas around Alaska to see where exploration is going on and which companies are involved. New deposits are described. In the current issue, recent gold discoveries in the Tintina and Kuskokwim gold belts are profiled. Placer mining activities, once a big emphasis in Alaska, are also included.

"Companies look at the placer data and it sometimes tells them where to go," Szumigala noted. "The mining activity helps the state grow. It's a local producer of jobs, and it helps the economy in the region. Alaska doesn't have a lot besides resources going for it in many places."

Before becoming Alaska's Senior Economic Geologist, Szumigala, or "Zoom" as many call him, had worked in many parts of Alaska. "Starting in 1983, I worked in the Aleutians, Eastern Interior, and the Southwest quite a bit. Depending on the era, I was looking for a Red Dog, a Greens Creek, or various types of gold deposits. I started in mineral exploration in 1978 looking for uranium. No one would look for uranium in the United States these days, but at the time it was a hot commodity."

A lanky geologist with a long-standing interest in granite, Szumigala has seen many changes in Alaskan exploration. "I did my doctorate on igneous rocks and gold in southwestern Alaska," he told me. "That was before we knew how big Donlin Creek was. I worked for Battle Mountain out there. I tried to synthesize all the information that was available. Donlin Creek was just being drilled. Now it's estimated to contain 11.4 million ounces of gold. Donlin Creek is Alaska's largest gold resource, but it's not mined, because of economic constraints. Right now they need a little higher grade, or one of the other factors has to change. Being the biggest isn't always the best. Being close to infrastructure helps a whole lot."

KEYS TO SUCCESS

Road access, power, existing mine processing facilities, and a labor force are key factors in reducing mine costs. Donlin Creek's remote location and low grade ore will keep it in "reserve" status. On the other extreme, the high grade of the remote Pogo Gold Deposit southeast of Fairbanks offsets the cost of developing it.

"The eastern interior of Alaska has a road system, railroad, and power," Szumigala pointed out. "More and more in exploration those are factors to consider. That's why the Seward Peninsula is good. It's easy to get around because there are roads. There are copper porphyries out in eastern Alaska; if there was a transportation system, they might be given a more serious look."

"Location, location, location," has long been cited in the success of any business. In addition to access and infrastructure, mineral oversupply in the market can also prevent new mines from opening.

OVERSUPPLY

"Look at all the money spent on Quartz Hill in southeast Alaska," Szumigala recalled. "Now molybdenum is dead, because so many copper deposits produce moly as a by–product. They can satisfy demand for decades to come. Quartz Hill is almost all moly. That deposit is being held on to by Cominco, and it will be mined some day. It's a classic example—if they went into production now, they would glut the market. You wait for a good price, and you dump it on the market, and there goes your price."

Located about fifty miles east of Ketchikan, Alaska, Quartz Hill is one of the world's largest known molybdenum deposits. From its discovery in 1974 until completion of a feasibility study in 1983, U.S. Borax saw changes in the price of molybdenum from a high of $30 a pound in 1979 to a low of $3 a pound in the early 1980s. The company sold their claims to Cominco in 1992.

I asked Szumigala about today's tin situation—another example of an oversupply. "The Seward Peninsula has good tin occurrences, but you can't compete with Malaysia and Brazil," he explained. "They've got so much placer tin that lode mines can't compete. In the U.S., Alaska has the major tin resources, but it's more expensive to mine the Seward Peninsula than Southeast Asia. The U.S. imports something like eighty-five percent of the tin it needs. If tin ever gets a better price, Malaysia can mine and process its lower cost placer deposits."

PLATINUM GROUP ELEMENTS/METALS' MARKET

In contrast, demand for the platinum group elements (PGE) has outstripped supply. New uses in automobile catalytic converters and high-memory computer chips are only some of the factors behind the higher prices.

"Because of the mandate to use alternative energy in cars and the need for PGEs in the current fuel cell technology, we're seeing a whole new array of uses for these metals," Szumigala told me. "Russia isn't producing. South Africa's Bushveld has been mined for quite a while. Another reason the Stillwater Mine in Montana is doing so well is because the palladium price is so high."

He flipped through a *California Mining Journal* to locate metal prices. "Platinum is at $500 an ounce," he noted. "Palladium is around $700. That's why some people are looking for these deposits. Alaska does have world class platinum deposits that just haven't been developed."

Two mining conferences I recently attended had sessions on the platinum group metals. Until a substitute is found, these metals will be in high demand. As exploration shifts to the search for PGEs, production will rise until the supply begins to satisfy the market demand. Perhaps new types of deposits will be found. Lithium, for example, which used to be mined only in pegmatites, is now being extracted from brines in Nevada.

GREENS CREEK MINE AND RED DOG MINE

The history of the Greens Creek Mine in southeastern Alaska illustrates how several factors determine a mine's success. The mine owners, Kennecott Minerals and Hecla Mining, discovered the deposit in 1974. They had to work through land closures and trades with the U.S. Forest Service. The presence of several minerals besides silver, including zinc, lead, copper, and gold has helped the mine survive low silver prices and avoid being dependent on just silver production.

"They're optimizing all those metals to get them out of the mill circuit," Szumigala explained. "That's a great thing about a poly-metallic deposit. You hope one of your metals will carry you through. If all of them are up, you're making a great profit."

The mine at Greens Creek started up in 1989 and continued until 1993 when it was temporarily shut down due to low metal prices. Negotiations were started with the U.S. Forest Service for subsurface exploration rights to larger areas around the mine.

"They started out with a smaller ore body," Szumigala continued. "When they closed, they decided to go out and find another deposit. It's a number of ore bodies that together make a great mine." He expanded on the Greens Creek example. "Don't concentrate on the one deposit you know," he warned. "Look for others across the hill. A lot of companies say a true headframe exploration program pays off the best, adding to your reserves. Ore deposits are anomalous, and it's possible that there are other anomalies around it."

When a world-class mineral occurrence such as the zinc-lead-silver Red Dog Deposit in northwest Alaska is discovered, exploration and development of other zinc ore bodies is negatively impacted. Cominco Alaska Incorporated began production in 1990 at Red Dog, and the mine now accounts for approximately seven percent of the world's mine-produced zinc. By-products germanium and indium are also recovered from Red Dog zinc concentrate at Cominco's smelter at Trail, British Columbia.

"Cominco has spent a lot of money looking for more Red Dog types of deposits," Szumigala stated. "The Red Dog exploration was in the '70s and '80s; right now they're making their profit. It's a difficult ore to process, but their production was ten percent higher in the past year."

"Next year instead of a million tons of concentrate, they'll produce 1.2 to 1.5 million tons," Swainbank told me. "These [smaller] three or five percent zinc deposits just aren't going to make it."

With increased production from the mill, new discoveries, and potential for more reserves, the Red Dog Mine has an estimated fifty years of operation left at present rates.

"Cominco geologists are looking in Canada, and asking, 'Can we develop this?'" Szumigala wryly noted. "They can't even compete within their own company because of Red Dog."

GOLD MINES

After thirty years of intense gold exploration and mine development, major gold mines continue to come into production. However, the difference between the current costs of producing an ounce of gold and the current market price of gold doesn't provide the high profits once seen in the industry. In addition, international bank policies and government decisions to sell gold reserves cause the price to decline.

Even so, gold is still the exploration target of choice, and gold is sought after by both small-scale prospectors and big companies. "The price of gold peaked at $800 an ounce in mid-January in 1980," Swainbank recalled. "And that lasted until the mid-'80s when people began looking for poly-metallics—gold, copper, lead, zinc, and silver."

Fashions in minerals change. "Sometimes it's hard to pick what will be hot," Szumigala observed. "No one would have picked diamonds in Canada, twenty years ago." Now diamond mines are being developed in the Northwest Territories.

SUPPLY AND DEMAND

What makes a mineral important? Strategic metals were once considered good investments because of their limited supply and their critical uses in industry in space exploration, and the military, but "space race" minerals–beryllium, cobalt, bismuth, niobium, tantalum, and others–are no longer in high demand. The use of minerals in weapons and solar power systems is in a slump, and even though some computer chips and semiconductors still use germanium, it has largely been replaced by silicon. Similarly, while some titanium is needed to build submarine hulls and large aircraft during peacetime, there are ample supplies.

The supply and uses of minerals is reviewed in the annual *USGS Minerals Yearbook*, and in the *United States Mineral Resources*, edited by Donald A. Brobst and Walden P. Pratt, and published by the USGS as *Professional Paper 820* (1973). I checked gallium in the 1996 edition of the *Minerals Yearbook, Volume I: Metals and Minerals*. It reported a price above $300 because of an increased market for it in the production of lasers, telecommunications devices, computers, and solar panels. Exploding usage of such items will keep gallium in demand for the foreseeable future.

The United States currently imports gallium from France, Canada, and Russia, suggesting that gallium might make a good prospecting target. However, the chapter on gallium, germanium, and indium in *U.S. Mineral Resources* points out that these minerals are produced as by-products from the processing of the zinc sulfide sphalerite, and that gallium and germanium are recovered from coal ash, flue dust, and stack gases in industrial plants that burn large quantities of coal. Although gallium ores aren't found in natural concentrations, the element is inexpensively produced from zinc ores and recovered from coal ash. Nonetheless, if an unusual mineral shows up in an assay report, the prospector should consult these references to see where possibilities exist for a market.

SUMMARY

Considering all the factors that influence the exploration for and development of a mineral deposit today, Dick Swainbank shook his head. "It's not like it was thirty years ago. Now, there's restricted access, different land status, and environmental restrictions. It used to be you could just go with the geology." Add to that list fluctuating demand for metals, substitutions, and recycling, and a prospector may well feel like a participant at a fashion show trying to figure out next year's dress length.

20. Black Hills Pegmatites

VALUE OF PEGMATITES

Few of the visitors watching the sun illuminate the four presidents' heads at Mount Rushmore understood the effect this granite had in creating world class pegmatites in the surrounding Black Hills of South Dakota. Named the Harney Peak Granite, this ten-mile diameter intrusive is really made up of many large and small granitic bodies. Multiple pulses of granite spawned a remarkable number of pegmatites, coarse grained pockets containing enormous crystals of beryl, amblygonite, spodumene, tantalum, cassiterite, and "books" of muscovite mica.

In many places the granite melt merged imperceptibly into coarse-grained pegmatites, while other pegmatites were emplaced in fractures in the schist. The Precambrian Harney Peak Granite and the still-older quartz-mica schists and quartzites were uplifted and exposed in the Black Hills dome of forested hills amidst younger sediments.

The town of Keystone began with small-scale gold mining and it now serves as the entrance to Mount Rushmore and Custer State Park. The famous Etta, Peerless, and Hugo pegmatites are visible from downtown and nearby are the Keystone, White Cap, and Bob Ingersoll pegmatites. Ten-foot crystals of spodumene, masses of purple lepidolite, sheets of mica, creamy feldspar for ceramics, and black crystals of cassiterite, niobium and tantalum were mined here.

During my trip, I found Black Hills rock shops and their owners were the key to understanding the rich pegmatite endowment of the area. At the Rock Shed, about a half-mile east of the tourist shops of Keystone next to Mineral Technology (Mintec) Corporation's processing plant for glass and quartz, Eugene Kuhnel has a small store filled with minerals and a yard overflowing with still more.

"We've had this shop for twenty-five years," he told me. "We've got rose quartz, mica, different minerals from the Black Hills, petrified wood, and rocks from all over the world inside the shop.

"There're a lot of famous mines around here, but no fee digging. Somebody should [start that]. The Etta is private land. It was an early mine for spodumene. I never mined there, but I sorted the dump for quartz. I worked for Mintec for ten years. We mined it for tantalum for a while. The Hugo is also closed. We mined it years ago for quartz and feldspar. I tell people it's all private land. However, there are a lot of roadcuts where it would be fine to collect samples."

As I was driving west after leaving Keystone, just past Mount Rushmore, the roadcuts revealed half a dozen pegmatites in the darker schists. Creamy feldspars caught my eye, and I stopped to sample these ribbons of lighter rocks. Most are simple pegmatites lacking multiple zones or a core of quartz. A few had muscovite mica and the promise of other minerals–if a person had time to trace them into the Forest Service land.

I continued on toward Hill City to visit James and Marilyn Dean's Dakota Stone Rock Shop, just two miles north of the city. Wearing dirty jeans and leather work gloves, the broad-shouldered owner was picking through chunks of rose quartz when I arrived.

"My Dad was in the feldspar mining business when I was in high school," he told me, and pulled a cold soda from a pop machine nearby. "He and another guy started this business in 1976, and after he died in a car crash, I bought out the rest of the family. I've been around pegmatites all my life."

Bins holding chunks of honey onyx, red feldspar, green quartzite, petrified wood, pudding stone, and purple and yellow lepidolite surrounded us in the yard. Sun glinted off the shiny surfaces of mica in many of the rocks.

"We were the last ones to work the Etta for quartz. We're currently working the White Cap for building material–schist for building stone–taking the high wall down. There's no beryl market, no mica market. [Even] the feldspar market is soft."

He paused to direct two fellows with a small front-end loader who were moving the rose quartz.

"We're mining rose quartz south of Custer at the Red Rose Quartz Mine, but our main business now is slates. We have four different quarries we're mining slate out of."

When he went off to supervise, I entered the store, noting that its exterior was faced with gray slate and the large fireplace inside had more of the same. The Dakota Stone had heaps of building stone and chunks of pegmatite material in the yard, and inside was a complete rock shop of delicate pegmatite minerals and crystals, agates, and displays of popular minerals and lapidary supplies.

Clustered around Hill City are tin-bearing pegmatites worked since the 1880s. The Cassiterite Lode, Tin Boom, Mohawk Tin, Tin Chance, Tin Key, Tin Spike and Cowboy Mines once supplied ore to the Harney Peak Tin Mining, Milling and Manufacturing Company near Hill City. However, most mines never shipped any ore, and now the tin boom is long gone. Locals no longer remember the location of these pegmatite mines.

The antique steam locomotive of the Black Hills Central Railroad was just taking on water when I arrived in Hill City. The train once carried tin ore, gold, and timber, but now it chuffs off to take passengers for a ride through the wooded back country. On the main street of Hill City I located the Black Hills Institute of Geological Research and the Black Hills Natural History Museum. The free museum has displays of local and worldwide minerals and fossils, and the staff specializes in dinosaurs. The gift shop has a complete selection of minerals, fossils and books for sale. Their website address is: www.bhigr.com.

Fifteen miles south of Hill City, more than one thousand five hundred pegmatites were mapped within thirteen square miles around the town of Custer. Pegmatites in this area are famous for their mica, lithium, beryllium, and feldspar. One of them, the Beecher Lode, had abundant amblygonite, spodumene crystals, and beryl. Another, the Helen Beryl, had disseminated beryl in green, yellow, white, and smoky crystals. Also nearby is the White Spar, one of the largest mica mines in the Black Hills.

ZONING IN PEGMATITES

Hundreds of the Black Hills pegmatites are made up of many zones, illustrating several stages in the slow cooling of silica-rich, mineralized fluids—the end products of crystallization of a watery granitic melt. Some geologists feel that pegmatites are a transitional feature between granites and quartz veins. Their principal minerals are quartz, the sodium feldspars microcline and albite, and muscovite mica. Tourmaline and apatite are also present in many pegmatites.

In complex pegmatites, several zones from the border to a quartz core vary in thickness and composition. Where the pegmatite contacts the schist, a thin border zone forms a fine-grained selvage of granitic and schistose material, scrap mica, and beryl. Inside the border zone, a wall zone may range from inches to feet in thickness, and is often one of the most productive zones for sheet and scrap mica, beryl, cassiterite, spodumene, niobium, and tantalum. Farther in, an intermediate zone of larger crystals forms the transition between the wall zone and the core. Here mica, beryl, and amblygonite are found. The core may be a continuous body or it may appear in several segments, depending on the thickness of the outer zones and the shape of the pegmatite. Quartz is usually the most abundant material, but it may be intermixed with lepidolite, microcline, spodumene, and cassiterite.

MICA

Sheet mica was the first pegmatite mineral mined, with more than eight thousand pounds produced from the Black Hills in 1880. Following a quiet period, Westinghouse Electric and Manufacturing Company bought up claims and reopened the New York, White Spar, and other mines around 1906 and worked them until 1911. Then, after another lull, the Colonial Mica Corporation purchased mica of "strategic" quality during World War II from more than two hundred and seventy-five mines and prospects here.

Although there are several types of mica, muscovite mica was the only commercial type mined from the Black Hills pegmatites. Muscovite got its name because Russians or Muscovites once mined it to use as window glass. These days, mica is mined in two forms—sheet and scrap.

Mica's flexibility, elasticity, toughness, nonconductivity, and colorlessness are the qualities most valued. Mica can be cut, punched, stamped, and machined to close tolerances. Better quality sheet mica is used in magnetos, spark plugs, vacuum tubes in radios, and condensers. Poorer quality mica is used as insulators in electric appliances, and for stove and furnace windows. "Scrap" mica is ground up to coat roofing and tar paper for use in wallboard, and decorative wall paper, as a filler in paint, in joint compounds and plastics, as a lubricant for industrial molds and in oil, or for coating tents and tarpaulins.

The New York pegmatite in the Custer District produced approximately two thousand tons of sheet mica, and from 1942 to 1945 the largest amount of sheet mica was hand-picked from wall zones in the Hugo and Dan Patch Mines. Now India provides ninety-five percent of the sheet mica used by the United States. However, sheet mica use has declined as synthetic materials have been developed. Vacuum tubes that required sheet mica have been largely replaced by transistors and printed silicon circuit boards.

In the Keystone District, the main scrap mica mines were the Bob Ingersoll, Hugo, Dan Patch, Peerless, and Wood Tin. Today, North Carolina is the leading domestic producer of scrap mica, with South Dakota about seventh in domestic production.

Just off Custer's main street is Pacer Corporation's mica processing plant, and nearby are their offices.

"We now process mica schist instead of muscovite mica," dark-haired Jeanine Gould, Pacer's customer service manager, informed me. "Our mica plant has been operating for twenty to twenty-five years. Most of our mica mines are within twelve miles of Custer. When we tested the mica schist, it had eighty percent mica content."

She showed me a dark chunk of the shiny mica schist they quarry just north of Custer.

"We grind it down to a specific size and export the finished product to companies in Malaysia, Taiwan, Korea, Peru, Canada, and Mexico. The biggest use is for fire-proof coatings. It's a mineral these countries just can't get otherwise."

FELDSPAR

Pacer Corporation also advertises itself as the "Home of Custer Feldspar" on its plant west of town. It produces potash feldspar from Black Hills pegmatites, but because of the distance to markets, feldspar was not mined here until 1923. Microcline, with its pale buff or cream color, and albite, in the form of cleavelandite, are the principal feldspars that have been produced from the Keystone, Hill, and Custer areas. The Hugo and Peerless Mines have some of the purest feldspars mined in the United States.

Gould told me that her company ships feldspar to Canada and to domestic manufacturers.

"Our feldspar plant has been operating for fifty years. Feldspar is used for anything ceramic-related: porcelain coatings on the cast iron tanks in hot water heaters, and for bathtubs, toilets, sinks, and chinaware. We also have a Wyoming deposit we selectively mine for dental grade feldspar."

PEGMATITE EXPERTS

About a half-mile east of Custer, I found Scott's Rock Shop and inside is a free museum, the "Rockhound Headquarters in South Dakota," and just down the road was Ken's Minerals and Trading Post. Scott's Rock Shop is open from March to December and is located at 1020 Mt. Rushmore Road, Custer, South Dakota 57730; (605) 673-4859. The yards in front of both places glinted with pegmatite minerals, rose quartz, and bins and flats of other chunky specimens. Earlier on the day I visited, Sam Scott had been out collecting apatite with his son and grandson.

"My grandfather and father started this shop in 1927, a year after I was born," he proudly informed me. Displayed on one wall is his grandfather's geological map of the Black Hills. The family has had several generations involved in rocks, including two sons who are professional geologists. Scott's Rock Shop has a separate room for displaying outstanding museum specimens, fossils, arrowheads, and trophy game heads. The large store has rooms for lapidary supplies, books, jewelry, agates and many, many minerals. They also sell fishing licenses, lures, and bait.

When I asked about pegmatites, people in Custer referred me to Vernon Stratton. I was fortunate to find him at his home near Custer finishing off a big breakfast with black coffee. Looking like the crew boss, cowboy and Homestake Mine lead driller he once was, he began a fascinating story about early pegmatite mining in the area.

"Every hillside had a pick and shovel hole in it. People would get an old Model A pickup, load it with some [feld]spar, and bring it into town. There was a time when there'd be fifty trucks lined up. If they could get a bucket of mica or haul in a ton of feldspar, they could get good money. Dad would bring in twenty-five tons a day with a little truck.

"The first beryl I sold, I got $.30 a pound. There was a buying station next to the grade school. I was in fourth grade. An abandoned garage had beryl stacked around it, and I'd pick up a chunk and sell it."

Beryl is found in almost all of the zoned pegmatites on the flanks of Harney Peak in the southern Black Hills. They are either within the granite or the surrounding metamorphic rocks. But beryl is often difficult to distinguish from quartz or feldspar because of its lack of crystal appearance and yellowish to white color. Even experienced hand sorters frequently missed beryl, so many pegmatite dumps are a good source for collecting.

"When I was six years old," Stratton reminisced, "my dad mined at the Calico Mine. I picked up beryl off the dump for him. They were getting $50 a ton. That was awful good money."

The Custer District has many mines and dumps with beryl. The most productive deposits are in pegmatites that have albite and muscovite. In his 1953 USGS *Professional Paper 247* on the area, Lincoln Page and others felt that the Helen Beryl pegmatite had the largest potential beryl reserves. In general, the length of the pegmatite and the productive zone determined the richness of the deposit. The width of the beryl-bearing zones was never more than five or six feet. One percent beryl was considered a rich deposit.

I asked Stratton about the Helen Beryl pegmatite west of Custer.

"My Dad initially mined it," he recalled. "He got up to $30 a ton for spodumene. It all went to Germany. It had hiddenite, a green, gem-quality spodumene. The wall zones had square crystals of tantalite containing fifty-five percent tantalum. I once sold a double square crystal. There was lots of small, gem-quality beryl, and a large amount of quartz. I mined the tailing piles and sold them in town. But the Forest Service bulldozed the mine. It had filled in with water, and they were worried about the liability."

BERYL

All of the dumps in the area had additional resources of beryl. However, none was rich enough to just produce beryl, so it was recovered as a by–product of mining mica, feldspar, or lithium. Hand-sorting and milling were the two methods used to concentrate beryl.

Beryl occurs as green and yellowish-green crystals and as white, pale pink, or blue crystals. Fine-grained beryl is found in simple pegmatites while more complex zoned pegmatites have larger crystals and masses.

Vernon Stratton mined pegmatites in South Dakota's Black Hills. Light-colored sodium feldspars, abundant muscovite mica, and large crystals are typical features of these deposits.

On a visit to the Tip Top Pegmatite, Stratton casually walked around identifying beryl. The yawning cavern of the mine was flooded, but the walls showed the massive crystalline structure common to pegmatites.

"I had a crew of eight working in the Tip Top in 1979," he explained. "We mined one hundred eighty-six tons of beryl and got $310,000 that year. We had a pretty large operation. I saved all the minerals. The Smithsonian got a flat of new minerals out of it."

Some of the new minerals discovered were tiptopite, fransoloite, pahasopaite, ehrleite, and tinsleyite.

While few beryl crystals from the Black Hills mines are considered gem quality, chunks of beryl from this area were an important source of beryllium for industrial use as an alloy with copper (seventy-five percent of beryllium's use), for parts in rockets, satellites, missiles, computers, and lasers. X-rays pass through pure beryllium, so it is used to make the small windows in X-ray tubes. Beryllium also has high electrical and thermal conductivity, and high strength and hardness when alloyed with copper or aluminum. However, beryllium dust can cause a serious lung disease, and early drilling and blasting during mining may have contributed to many deaths.

Before taking me to the beryl-rich Beecher pegmatite, Stratton first checked in with its owner Joyce Bland.

"Always rocks, rocks, rocks," she said of her deceased husband, George. "He was really in love with beryl and he ignored the spar."

"I cobbed spar here," Stratton interjected. "I was eight, ten years old. My Dad worked out here. George had six employees mining beryl, and a processing plant. He sent out train loads."

We hiked over to the rusty walls of a deep ravine where the pegmatite was located. Stratton led me down the crumbling slopes to peer into a tunnel.

"So much iron in it, the feldspar wasn't saleable," he told me. "All through this mine there are massive zones of mica. Pacer needed some mica a few years ago, so we came out here and picked out the bigger books. Some zones run twenty to thirty percent mica."

OTHER MINERALS

The important lithium minerals, spodumene, amblygonite, and lepidolite, are formed late in pegmatite genesis. Intermediate and core zones of complex pegmatites contain the lithium minerals, surrounded by barren outer zones. Lithium-bearing pegmatites are found in all areas of the Black Hills, with crystals being found in the larger deposits that have well-defined zoning.

West of Custer, Stratton directed me to the privately owned Tin Mountain Mine nestled in the lodgepole pines. Outside the excavated cavern were boulders of spodumene and amblygonite. An old blackened campfire indicated another use for pegmatite workings.

"It was mined for cesium, lithium, and tantalum," he said, and pointed to a black splotch on the ceiling. "There's a big tantalum crystal they didn't reach," he noted. "The dump is fairly rich in tantalum." He traced white logs of spodumene along one wall. "And there were four-foot beryl crystals," he indicated with outstretched arms. "This was definitely one of the unique mines of the Black Hills."

Spodumene crystals are usually white or gray, but green spodumene (hiddenite) was found in the Etta Mine and pink (kunzite) in the Beecher Lode. Crystals ranged in size from grains, to laths up to forty-seven feet long, as in the Etta. Many crystals were one to ten feet in length. Surface leaching frequently created "rotten" spodumene by replacing the rock with micaceous minerals, which reduced the lithium content.

Amblygonite is usually white or gray to colorless rounded masses, from one inch to ten feet in size. Amblygonite bodies are often covered with a thin layer of muscovite, and crystal faces are rarely seen.

Lilac, pink, violet-gray, and honey-yellow lepidolite, a lithium-bearing mica, occurs as massive aggregates of small crystals or as scattered flakes. They are often associated with pink tourmaline, blue beryl, cleavelandite, quartz, and microlite. Most lithium-bearing pegmatites are large, irregular bodies, such as in the Beecher, Etta, and Bob Ingersoll pegmatites. The Bob Ingersoll was the largest source of lepidolite in the Black Hills.

Ceramics and glass manufacturing processes are the largest consumers of lithium compounds in the United States. Thermal shock-resistant cookware is a major use of lithium. Other uses include lithium batteries and lubricants, aircraft alloys with aluminum, and pharmaceutical-grade lithium for treating manic-depressive psychosis.

"George Bland was sure that the lithium water from the Beecher pegmatite made his sons mellow," Stratton joked. "That and the goat's milk."

There are more than twenty-five spodumene-bearing pegmatites in the Black Hills. While spodumene is the most common lithium ore, extracting lithium requires an energy-intensive chemical recovery process. Since 1966 Cyprus Foote, the world's largest lithium company, has produced low-cost lithium by evaporating lithium-rich brines at Silver Peak, Nevada.

At Carl Scott's Red Rose Quartz Mine, hidden off Forest Service roads near Custer, Scott Kellum took Stratton and me up to the rubble-filled opening.

"We find a lot of bright red rose quartz," Kellum told me. "We supply Jim Dean's Dakota Stone in Hill City, Scott's Rock Shop in Custer, and rock shops throughout the United States."

They were reworking the old dumps and expanding the working face. Near the mine entrance, big cribs held tons of rose quartz.

"We grade out the rock as we mine it," Kellum said. "We sell two or three different grades."

Two miles east of Custer, past Scott's and Ken's Rock Shops, the Shamrock pegmatite is visible from the highway. Once known for gigantic pods of pure feldspar, the workings are now filled with water. The walls of the Shamrock have the typical light pegmatite coloration from creamy feldspars and muscovite mica. Adjacent areas that contain watermelon tourmaline and rose quartz have been filled in.

"This country hasn't been touched yet," Stratton asserted. "People go out to mine surface feldspar. In the Southern Hills beryl and tantalum are deeper. I could put two hundred people to work tomorrow, if it was like the 1950s and 1960s, before the Forest Service and government regulations."

FOREST SERVICE LANDS

I met with Don Murray, the Black Hills National Forest land and minerals person, in a quiet Forest Service office near Deadwood, South Dakota.

"Most of the pegmatites in the Black Hills are under claim," he cautioned me. Pacer is operating a lot of them. We don't have a place set up for rockhounding yet."

He selected his words carefully, trying to educate me and others who might come to dig for minerals.

"I don't know of any Forest Service land where people can't go and grab a rock sample for their own pleasure. But you do have to deal with the mining claim issue. There are a lot less claims since the annual federal claim rental fee of $100 started in 1993. We had around seven thousand claims, but now it's probably about half that.

"I get fifteen to twenty operating plans for pegmatites a year—some existing, some new. Deposits are always being reclaimed and reopened.

"Then, there's the abandoned mine issue. Some sites are hazardous to the public because of unstable overhangs and deep water. While we are starting to do reclamation, we are also looking for a site for rockhounds so we can open it up for recreational use."

SUMMARY

When I left the southern Black Hills after my brief tour of several pegmatites and rock shops, the car floor was covered with mica flakes, and clay from decomposed feldspar, evidence that in addition to its wooded hills, fish and game, lakes, and campgrounds, the area hosts spectacular mineral occurrences. On the highway south of Custer, some easily accessible roadcut pegmatites caused me to swerve across the divider for a closer look, and considerably delayed my departure from this mineral-rich area. While many of the more famous localities are private, the rock shops and roadcuts still yield satisfying treasures.

21. THE DARK ROCKS AND PLATINUM METALS

WHERE TO FIND PLATINUM METALS

Platinum metals are found in economic quantities in only a few areas of the world. Best-known are the placer deposits in Colombia, Alaska, Russia's Ural Mountains, and lode deposits in South Africa, Siberia, Canada's Sudbury area of Ontario, and Montana's Stillwater Complex northeast of Yellowstone Park. The unique qualities and the scarcity of platinum group elements (PGEs) have resulted in a large amount of literature about their geology.

On a spring afternoon in May, I drove up the lush valley of the Stillwater River in Montana. I stopped on Absaroka's main street to talk with two old-timers in front of Absaroka Cinch, where saddle accessories are made.

"I worked at the Crow Mine," Gordon Campbell told me, gesturing toward the Beartooth Mountains in the distance. "My Dad worked there, too. It was a chromite mine. I worked in winter putting in track out to the waste dump. It got really cold and there was a lot of snow. It was a treacherous job. Everything was straight up and down."

"There's a pile of chromite near here, at Columbus," Everett Imlay added. His cap had a crossed pick and shovel. "It wasn't worth anything after the Korean War, and they just quit mining it."

Discovery of the platinum-enriched layer in these mountains resulted from good basic research, prospecting, and improvements in assaying methods. In 1967 the Johns-Manville Corporation started a field exploration program in Montana's Beartooth Mountains. The previous mining of chromite there, and the occurrence of nickel-copper sulfide zones hinted at similarities to the platinum-rich Bushveld Complex of South Africa. A decades-old mention of platinum in a geology report also directed attention to the area. Coincidentally, in 1968 new analytical methods of assaying had improved testing for platinum and palladium in soils.

Discovery of soil enriched in the PGEs, mineralized boulder float, and magnetic and electrical responses in the bedrock eventually led to the Johns-Manville Reef. Underground exploration tunneling began in 1974 to 1975 to establish the width, grade, and extent of the zone, and the results showed the Johns-Manville Reef was a major resource of palladium and platinum, far richer than the Bushveld Complex, though it's more limited in extent.

Near the head of the valley, above the clear-running Stillwater River, the mill and offices are tucked against the steep mountain face. Approximately sixty percent of the ore is transported to this mill via a vertical shaft near the building. The rest of the ore is moved by rail or truck to the mill.

"The nearby town of Nye had a population of 4,000 in the 1940s and 1950s," Chris Allen, one of Stillwater's Mine administrators, explained. "That's when the government operated a chrome mine for its strategic stockpile. Now, most of our employees live in Fishtail, Absaroka, Columbus, or Red Lodge."

The Stillwater Complex was originally mined for its chromite layers, which were as thick as three feet. In a different layer, traced throughout the Stillwater Complex, PGEs in the form of sulfides, arsenides, tellurides, and alloys are associated with the disseminated copper-nickel-iron sulfides of chalcopyrite, pentlandite, and pyrrhotite.

The Stillwater Mine near Nye, Montana produces the only domestic platinum and palladium in the United States. An ultramafic layered intrusion twenty-eight miles long contains chromitite and the platinum group elements.

PLATINUM ELEMENTS

It wasn't until the nineteenth century that Europeans discovered that platinum was actually six separate elements. This group of often-alloyed elements includes platinum, palladium, iridium, osmium, rhodium, and ruthenium. William Wollaston, a London chemist, separated platinum from its sister elements and discovered palladium in 1802 and rhodium in 1804.

All PGEs have a high melting point (2820.20°F to 4892°F or 1549°C to 2700°C) and resist most acids. Sperrylite, the most widely distributed platinum mineral, contains platinum, arsenic, and minor rhodium. Iridium, with its specific gravity of 22.65, is the heaviest of the platinum group elements found in nature (gold has a specific gravity of 15 to 19, and platinum is 14 to 19). Only gold and silver are more malleable than platinum. While platinum is considered a precious metal, more than eighty percent of its use in U.S. is for industrial purposes.

DARK IGNEOUS ROCKS

Chromite, PGEs, copper, nickel, and magnetite are most often found in the dark igneous rocks—dunites, gabbros, and peridotites and their altered equivalent, serpentinite. The dark green silicates—pyroxene, hornblende, olivine, and the clear-colored plagioclase—give these rocks a characteristic black, dark green, brown, bronze, or splotchy mottled color.

Known as mafic and ultramafic igneous rocks because of their high magnesium and iron (ferric) contents, these dark igneous rocks originate either from oceanic lithosphere, which may be obducted onto a continent and become an ophiolite, or as isolated, circular intrusions in an ocean island or along continental margin settings. Igneous rocks having eighty percent or more mafic minerals are identified as ultramafic. Pyroxenites have greater than sixty percent pyroxene; peridotites have more than forty percent olivine; and dunites have ninety to one hundred percent olivine.

Unlike the tectonically sliced and diced ophiolites, the large, saucer-shaped mafic to ultramafic magma chambers, which were emplaced in a less active tectonic environment, have minerals in distinctive layers. Some writers characterize these layered complexes as being similar to a sedimentary settling out process that took place in a heated chamber over a long period of time. Eventually, chromitite, segregations of iron-nickel-copper sulfides, and platinum minerals coalesce as layers. Then, uplift and erosion brought slabs of these mantle rocks to the surface. Originally flat-lying, some were tilted on end, as seen in the Stillwater, Montana rocks.

PLATINUM DEPOSITS

Platinum was initially mined from placers, where it occurred as dust and flattened pellets. About one-third of the world's platinum still comes from placers. The Spanish discovered that Ecuadorian Indians were using platinum and gold for jewelry, and they named platinum "little silver," or platina, because of its silver-white color. Spaniards used indigenous people to work the Colombian placers beginning in 1778, and these sources continue to supply a small part of the world demand.

Ultramafic rocks on display in an ore car outside the Stillwater Mine contain the dark minerals hornblende, olivine, and pyroxene. Ultramafic rocks originate in the mantle and oceanic lithosphere.

By far the richest placers are located in Russia's Ural Mountains. These were discovered in the 1820s and they produced huge nuggets, some over twenty troy pounds. Paystreaks were extensive; one in the Iss River extended fifty miles to the Tura River where it continued another fifty miles! The headwaters contained large nuggets, while far downstream the platinum was very fine-grained and difficult to recover.

With its wealth of platinum, Russia minted platinum three-, six- and twelve-ruble coins from 1828 to 1841. The three-ruble coin was the size of a quarter and contained ninety-six percent platinum. These coins are rare and bring high prices today. Some Russian platinum and palladium now comes as a by–product from Siberia's Noril'sk copper-nickel ores, in which palladium is two to ten times as plentiful as platinum, making Russia the leader in world palladium production. Other Russian sources of PGEs include the Pechenga Nickel Mines and placer deposits in the Russian Far East.

A modest placer deposit was discovered along Alaska's western coast at Goodnews Bay in 1926 by Walter Smith and Henry Wuya, two Eskimos familiar with prospecting. The Goodnews Bay Mining Company acquired the platinum-bearing creeks in the area, built a Yuba dredge on the property in 1937, and was the sole operating company after 1940. More than 500,000 ounces of platinum were recovered, but large scale dredging of the deposit has ceased, even though the lower grade paystreaks have not been worked out. The largest nugget recovered weighed four troy ounces.

Nearby the Red and Susie Mountains have dunite cores–an igneous rock that weathers to a dun color–and outer zones of peridotite, clinopyroxenite and hornblendite. Some of the dunite and peridotite making up these two mountains has been altered to serpentinite. The dunite, peridotite, and serpentinite contain disseminated chromite and small amounts of platinum and are considered to be the sources of the placers. During placer operations, platinum is often found intergrown with chromite, suggesting a common origin.

Jeff Foley, formerly a U. S. Bureau of Mines Alaskan geologist, now works with Calista Corporation, an Alaska Native regional corporation that controls the hardrock resources of the Goodnews Bay Platinum Deposit.

"The placers were bought from the Goodnews Bay Mining Company by Raymond Hanson of Spokane, Washington," Foley told me. "Corral Creek Exploration of Denver, Colorado has an exploration and mining lease with Calista for the Red and Susie Mountain area." So far, their work has involved soil sampling, mapping, and geophysical work.

"Most platinum group metal placers," he explained, "are derived by erosion of dunites and peridotites, whereas layered igneous complexes such as those at the Bushveld, Stillwater, and Russia's Noril'sk do not tend to form placer deposits. Most of the world's platinum is associated with either chromite or nickel sulfides."

The Red and Susie Mountains represent a disseminated, low-grade type of platinum source that forms rich placer deposits, similar to the Russian Ural Mountain placers. Montana's Stillwater complex represents the other type of deposit in which the platinum metals constitute a better lode. Because of their mixture with sulfides, the platinum metals in a Stillwater-style deposit break down in the oxidizing stream environment and don't form economic placers.

"The layered igneous complexes found at Stillwater, the Bushveld, and Noril'sk tend to have very complex PGE mineralogy," Foley told me. "They've got arsenides, tellurides, bismuthides, antimonides, and sulfides. You've got a host of minerals, including up to seventy-five mineral species containing PGE minerals."

The layered rocks found in a mafic-ultramafic magma chamber may contain both oxides (chromite, magnetite, ilmenite, cassiterite, and rutile) and sulfides (pentlandite, pyrrhotite, chalcopyrite, pyrite, and sperrylite) that illustrate a differentiation process–a settling or unmixing of incompatible (or "immiscible") minerals. Only under the high pressures and temperatures of the magma chamber could this mixing initially exist. Then, layers may form based either on mineral content, such as silicates, oxides, and sulfides, or density–grading upward from dense to less dense (dark- to light-colored) minerals or grain size, trending from coarse- to fine-grain size, as the melt crystallizes.

Foley explained that during magmatic differentiation, PGEs concentrate either in sulfides or oxides (chromite and magnetite), depending the magma chemistry, temperature, and pressure. Ultramafic rocks in both zoned and ophiolite complexes tend to be low in sulfur and sulfides. Where sulfides predominate, PGEs combine with sulfur, arsenic, tellurium, and bismuth. Where the oxides chromite and magnetite form, the dominant PGE mineral is likely to be ferroan platinum or an osmium-iridium alloy.

"Platinum placer deposits are composed of two different mineral species," Foley explained. "They are iron-platinum alloys, or ferroplatinum, and alloys of PGE with osmium and iridium as major components of the alloy. Depending on where you are in the world, one or the other may be more common.

"The Ural Mountains, Goodnews Bay, Colombia, and Ecuador placers have nearby lodes containing disseminated platinum and chromite in layered dunites and peridotite complexes. The platinum iron alloy is the most common PGE in these placers."

"In contrast, ophiolite complexes [slices of oceanic lithosphere containing mafic and ultramafic igneous rocks and sedimentary rocks] tend to be enriched in osmium and iridium relative to platinum. Sulfide minerals are less common than chromite and magnetite in the ultramafic portions of ophiolites and zoned complexes."

South Africa produces seventy-five percent of the world's platinum and Russia is second with about twenty-five percent. The Stillwater Mine in Montana and the by–product platinum that is obtained from United States copper and gold refineries provides only about ten percent of the United States' needs, so we are heavily dependent on imports from South Africa and Russia. In 1997, palladium cost about half as much as platinum, but now its price exceeds platinum.

The world's largest layered complex, the Bushveld Igneous Complex (BIC), located in South Africa, contains eighty percent of the world's platinum reserves. This storehouse of minerals has the world's largest deposits of chromite, PGE, vanadium and magnetite. The complex, which is about five miles thick and underlies an area of 15,000 square miles in four major lobes, extends 270 miles east to west, and 180 miles north to south.

The important platinum minerals occur in a one-yard thick seam named the Merensky Reef. The ore consists of pyroxenite with chromite, microscopic platinum in the form of sperrylite and ferroplatinum, magnetite, and specks of pyrrhotite, pentlandite, chalcopyrite, and nickeliferous pyrite. Sandwiched between chromite seams, this layer has a pegmatitic texture with very large pyroxene crystals. The overall effect is green crystals in a matrix of clear to light green silicates extending for miles.

The South African Bophuthatswana Mines produce seventy to eighty tons of platinum and palladium every year. New mines are increasing production from this area, but as these mines follow the layer deeper, they are hotter and mining costs are becoming more expensive.

Platinum metals are also produced as a by–product from massive copper-nickel sulfide bodies in Canada's Sudbury region of Ontario. Here, a huge meteor impact created the basin around which are found the productive mines. Geologic and mineralogical similarities to the Bushveld suggest a large meteor impact origin for both of these deposits.

STILLWATER DEPOSIT

Just off Interstate Highway 90 in Columbus, Montana, the Stillwater Mining Company has its smelter and base metals refinery down the road from the stockpiled black mountain of chromite and about thirty miles from the mine and mill in the mountains.

"The mill crushes the rock to extract the ore," Wendy Yang, Stillwater's Investor Relations Manager, explained. "It's concentrated to about sixty ounces per ton of combined palladium and platinum. We produce three and two-tenths ounces of palladium to one ounce of platinum–about seventy-five percent palladium and twenty-five percent platinum."

Concentrate from the mine is trucked to the smelter and base metals refinery in Columbus, and is used to produce "matte," a mixture of metal and sulfides, containing six hundred ounces of palladium and platinum per ton. Since 1990, furnace capacity has been thirty-two tons a day, but a new hundred-ton-per-day furnace is under construction. From the outside the smelter looks like a railroad industrial warehouse, with a guard station to discourage casual visitors. In addition to ore, up to three tons of old automobile catalytic converters can be processed each day at the smelter and their platinum and palladium recovered.

"Recycling automobile catalysts is an extremely small part of what we do," Yang told me. "We're primarily a producer."

The refinery produces "filter cake," composed of sixty percent palladium and platinum. Yang described the filter cake as looking like coal dust. "That's the form we ship from our base metals refinery to precious metal refineries for conversion into a finished product. We also get by–product rhodium, gold, nickel, copper, and silver."

At the end of 1998, the refinery shipped its one millionth ounce of palladium and platinum–and it's only been in operation since May 1996. Refinery capacity is also being increased from 4,000 tons per day to 7,000 tons per day. This is due to an increased output from the Stillwater Mine and their new East Boulder Mine coming on line in the northern segment of the complex about thirteen air miles away.

"We have two tunnel boring machines cutting fifteen-foot diameter adits [tunnels] there," Yang explained. "We expect East Boulder to be mining approximately 5,000 ounces a year by 2001. At our current rate, we are only four to five percent of current palladium market. Even after we grow to over a million ounces a year, that's less than ten percent of the world supply."

With about one thousand employees, the company expects to produce between 525,000 and 575,000 ounces of combined palladium and platinum by 2000.

As in the Bushveld, Montana's Stillwater Complex is a layered mafic-ultramafic sheet turned up on its side, with a maximum width of three miles and a length of twenty-eight miles. The Johns-Manville Reef varies randomly from several inches to seventy feet in width, both vertically and horizontally, along its length. The average ore width in 1998 was ten feet, and the grade was close to 0.7 of an ounce of PGE per ton of rock.

"We have fifty years of proven and probable reserves," Yang said, warming to her role in investor relations. "The deposit is open at depth. We've traced it 2,000 feet below our current working level. It has 'Blue Sky' potential, enabling us to triple our production in the next few years. We have the highest grade deposit and we're the only primary palladium producer in the western hemisphere."

I was impressed with the pristine setting of the mine near the Stillwater River, "a blue ribbon trout stream." From the dispatcher's office, I could see an area across the river being revegetated.

"The sprinklers are on former chrome tailings," Yang proudly told me. "The tailings left by the previous operators covered a big, gray area. We capped the tailings, reseeded it, and we are using mine water to grow a perennial grass, Garrison's creeping foxtail."

Above the mill is a holding pond where material is stored before being recycled into excavated areas in the mine.

"We began mining in 1986," Yang said, "And we've never had an environmental citation. We are very sensitive to preserving the quality of life. The disturbed area from the mill is about one hundred fifty acres—that's smaller than most malls."

The Stillwater Mining Company based in Denver, Colorado became a public company in 1994, and information about the company is available on their website at www.stillwatermining.com.

PLATINUM USES

Because platinum is inert and doesn't corrode or tarnish, it is used in jewelry for gemstone settings, rings, necklaces and pendants. Consumers in Japan, China, Germany, Italy, Switzerland, and North America are buying more and more of the white metal in necklaces, bracelets, bridal rings, watches, and earrings. Hong Kong entrepreneurs have set up factories for sales in northern China. Simple platinum jewelry is seen as sophisticated and elegant by younger buyers.

The major use of platinum, rhodium, and palladium is in auto and truck catalytic converters to reduce vehicle exhaust emissions. One ounce of these PGEs supplies enough material for coating the ceramic honeycombs of sixty-four auto catalytic converters. These inert metals transform carbon monoxide, nitrogen oxides, and hydrocarbons to harmless water vapor and carbon dioxide while themselves remaining unchanged.

New vehicle fuel cell prototypes using significant amounts of platinum are being designed, and petroleum refineries and electronics also use PGEs. Computer disk drives have a platinum-cobalt layer to improve the disk's ability to store large amounts of information. The glass industry uses platinum, rhodium, and palladium for making high quality computer monitors and liquid crystal displays (LCDs). Demand for these metals in most areas is increasing.

Platinum prices have ranged from $300 an ounce to more than $1,000 in 1980. Palladium prices were around $150 an ounce in the early- to mid-1990s. Its use is increasing in new automobile and wood stove catalytic converters, and the limited supply has caused the price to rise above $250 an ounce. As a result, South Africa and Australia are developing new mines, and Stillwater and Russia have increased their output.

SUMMARY

Consistently priced higher than gold, platinum appears to be the current investment metal of choice. The United States began minting platinum American Eagles in 1997 in four sizes: one ounce, half-ounce, quarter-ounce, and one-tenth-ounce. In addition, industrial demand for PGEs continues to be slightly more than the supply, assuring a strong market for these unique metals.

Although the gabbros, dunites, peridotites, and serpentinites are still somewhat mysterious creations from the deep ocean mantle, they bring us the platinum metals to clean our auto emissions, improve our electronic world, and gleam on sophisticated necks and fingers. How many gold panners have swept little dark grains out of their pans without discovering their value? Attention must be paid. These modern metals have a growing importance in our world.

22. Geophysical Prospecting

A physicist was asked to describe a cow.

"Think of the cow as a sphere," the physicist began.

In much the same way, geophysics looks at the physical properties of rocks in ways a hand lens cannot. Their magnetism, conductivity, radioactivity, fluorescence, and density are some of the more common qualities. Think of the earth as a sphere. Beneath its skin are rocks with differing magnetic properties and resistivity. Faults and buried ore deposits may be mapped without blindly drilling on a guess.

Geophysical methods are relatively new tools for examining rock properties and the hidden geology. Metal detectors, ultraviolet lamps, scintillometers (more sensitive than a Geiger counter), magnetometers, and portable electromagnetic systems provide prospectors with information for assessing ground for a potential mineral deposit. Since the 1950s the majority of Canadian deposits have been found using geophysical methods. Oil companies have regularly used seismic methods to map subsurface layers to find oil traps.

Not all geophysical methods require complicated equipment. I once used a metal detector to analyze the bare ground under a grove of aspen trees, listening to faint "null" responses. Digging up a small piece of soil, I crumbled it into my plastic gold pan and passed the detector's coil over it to verify the presence of iron. In this way I collected a thimble of flattened, gray metal pieces for my efforts. This was a form of geophysical prospecting, an attempt to learn why this particular area had given me a high magnetic response when I walked over it with a magnetometer, a more sophisticated instrument that reaches much deeper into the ground than a metal detector.

"Bench testing" with gold, silver, copper, or magnetic filings will show the amount and type of response a metal detector has to low- and high-grade ores, but sulfides such as iron pyrites are not conductive and

The author takes readings using a magnetometer. A sensor on top of the pole measures changes in rock types, faults, and concentrations of magnetite. The instrument gives a reading when a button is pressed.

they will not produce a response. Non-conductive magnetite and pyrrhotite, another magnetic iron mineral, cause a negative signal. On that hillside survey, the null signal I was picking up came from pyrrhotite weathering out of the decomposing schist.

METAL DETECTORS

Electronic prospecting with a metal detector, popular for nugget hunting, reaches through gravels and bedrock up to a foot or more, depending on the concentration of the minerals. Metal detectors are useful in searching for "hot" rocks in float on a hillside or for veins in a mine. These instruments create an electromagnetic field with their search coil that identifies conductors: gold, silver, copper, and other non-ferrous metals. I've used a metal detector to identify magnetite-rich rocks, to trace float from a vein, and for nugget hunting. Metal detectors can be rented for a short project or purchased.

Below the hillside I was examining with a metal detector, a small granite that was colored by the minerals it contained jutted out of the hillside. I suspected that the granite had formed a blind deposit deep in the hillside because conventional models of ore deposits point to granite as a potential source of minerals. The enclosing country rock was a soft greenish schist that crumbled into shiny mica (it had probably been a basalt before it was metamorphosed). Assays of soil samples from the slopes indicated that nothing was there, but my magnetometer work had found a circular ten-acre magnetic high. As another way to analyze the ground, I employed a metal detector to comb the surface for clues. It was giving me the answer, but, as usual, I wasn't receptive.

DRILLING

The next step was to drill. This costly and time-consuming experience eventually taught me to avoid drilling until all other options are exhausted.

"I've got an old rotary drill you can use," my prospecting teacher offered. "You'll have to get it running and drag it to the site. There's an equally ancient air compressor to go with it."

Some time later, after borrowing more parts to make the drill run, and paying for shop time to rebuild the compressor, I had everything hauled out to the site. A small CAT leveled a drill pad. I set up the derrick, stacked drill sections nearby and began to drill. No need to go into detail over the breakdowns, delays, near fatal accidents, and clouds of dust I generated.

When the borrowed drill didn't get deep enough for me, I hired a professional driller, brought in a lumbering giant of a rig, and filled hundreds of sample bags with the greenish schist. I finally put a magnet to the cuttings and pulled out more of the flat gray pieces of metal. It dawned on me that I had drilled a bed of metamorphic rocks with pyrrhotite, a magnetic iron mineral that my surface work with the metal detector had already suggested.

INTERPRETING THE CLUES

From this costly experience, I learned to thoroughly examine surface clues because they may hold the key to what lies below. Relatively inexpensive assays of soil samples should show some mineral leakage from deeper sources or along faults, particularly if there is some topographic exposure of the bedding. In my case, I ignored the unfavorable soil assays, feeling that there just had to be something there because of the nearby granite. After the magnetometer showed me a "hot spot" or magnetic high, the metal detector follow-up gave me conclusive evidence of an iron mineral, which I ignored. It was an expensive and time-consuming reminder that the geologic record is very complex, and full of such blind leads and false anomalies.

Geophysical information adds complexity to what often appears to be a featureless environment, and expands our knowledge of the geology. Interpretation of geophysical data, as with a crystal ball, may reflect what a person wants to see rather than the reality of what is there. Geophysical data and methods may suggest new targets to explore, from a regional to a local scale. The prospector's job is to understand the geology of the different rock units and identify geophysical anomalies–faults, favorable rock units, and intrusives.

ULTRAVIOLET LIGHT

At the micro-scale, I use a ultraviolet light on my dried pan concentrate samples to check for scheelite, a calcium tungstate, which readily fluoresces white to blue under a ultraviolet lamp. Scheelite occurs in several types of gold deposits as well as in tungsten skarns, and is resistant enough to form a placer deposit. The famous Carlin, Cortez, Getchell, and Gold Acres Lode Gold Deposits of Nevada averaged thirty-five parts per million scheelite.

Many other minerals, such as calcite, fluorite, zirconium, willemite, diamond, and ruby corundum, also respond to long- or short-wave ultraviolet light. When out in the field, a battery-powered ultraviolet lamp and a blanket (to create a dark "tent") can be useful, but I usually bring my rock and pan concentrates home, dry them, and use a plugged-in ultraviolet lamp to check for scheelite. However, beware of the white specks showing up under a ultraviolet light because plain old household dust also fluoresces!

SCINTILLOMETERS, GEIGER COUNTERS AND RADIOMETRIC SURVEYS

Since scintillometers measure alpha, beta, and gamma particles, while Geiger counters only measure alpha particles, they are more useful in geophysical prospecting for uranium, thorium, and potassium. Alpha and beta particles only penetrate earth materials a few millimeters, while gamma radiation is capable of penetrating two to three meters of solid rock. Uranium may not be high on everyone's list of minerals to seek out, but radioactivity can also lead to gem-bearing pegmatites, rare earth granites, and phosphate beds. Since granites have a higher amount of potassium than most sedimentary rocks, defining the extent of a granite by using a radiometric survey, may yield targets for copper, tin, uranium, or molybdenum.

New highly enriched uranium deposits in northern Saskatchewan, Canada, were discovered using airborne radiometric surveys. The source of the uranium was then traced by using ground scintillometer surveys. Ground surveys can be made while driving a Jeep, four-wheeler, or hiking with the instrument. Shallow placer channels containing minerals with uranium and thorium concentrations (zircon, sphene, apatite, xenotime, thorite, and monazite) can be traced with a ground survey with deeper channels being drilled and the cuttings tested for radioactivity. Faults, another structural feature prospectors want to identify, may have a radioactive high from migrating radon. The only limitations in scintillometer measurement occur from water-saturated ground and deep snow cover.

I've been trying to trace a vein of silver that cuts through a granite. Soft, crumbly granite near the vein shows alteration of the feldspars. Slope rubble covers the vein, but I should be able to trace the higher radioactivity of the altered zone with a scintillometer. If the silver is concentrated enough, a metal detector should pick up the location of enriched float along the hillside, because of its electromagnetic search coil. Combining these two methods, I should be able to locate the trend of the vein, and thus make it easier to expose.

GEOPHYSICAL MAPS

I visited Laurel Burns, a geologist and geophysicist for the Alaska Division of Geological and Geophysical Surveys (DGGS) in Fairbanks, Alaska, to learn more about the new geophysical maps the state is publishing. The DGGS has been doing detailed airborne magnetic and electromagnetic (EM) maps of mineral districts since 1993. For a list and prices, write:

STATE OF ALASKA, DEPARTMENT OF NATURAL RESOURCES
Division of Geological & Geophysical Surveys
794 University Avenue, Suite 200
Fairbanks, Alaska 99709-3645
(907) 451-5010 / (907) 451-5050 Fax
http://www.dggs.dnr.state.ak.us/

"The impetus is to encourage exploration," Burns explained from her curled position in her chair. "We're doing one-quarter mile line spacing for the flights."

These inexpensive magnetic and electromagnetic maps show amoeba-shaped splotches of yellow, orange, and red indicating the magnetic highs and conductive rocks, and green, purple, and blue for the magnetic lows and resistively-low rocks. These maps are also sold as topographic map overlays with the magnetic highs and lows shown as contoured lines, and electromagnetic anomalies as darkened circles. The data is available on a CD, and Burns demonstrated on her computer how the information could be programmed to emphasize particular features.

After the maps of the Fairbanks Mining District were published, claim staking shot up. Anyone could purchase these maps and find a target to investigate. Such geophysical maps of the well-known mining districts of Fairbanks, Livengood, Nome, Circle, Rampart-Manley, Ruby, and the Upper Koyukuk River in the Brooks Range have greatly increased the exploration activity in those areas.

"In addition, we make a new geologic map–given sufficient funding–after we do an airborne geophysical survey," Burns said, and led me to the publications room. "We revised the geologic maps of the Circle

Mining District with the new information by doing follow-up geologic fieldwork. The geophysics showed the field geologists where to go to find the edge of the pluton."

She pulled out the old geologic map and the revised one for comparison. "The geophysics enabled them to locate the boundaries of the plutons and a dense system of closely-spaced, high-angle, northeast trending faults," she explained. "The faults were visible on the aeromagnetic map because hot fluids migrated through the fault zones and altered magnetite to hematite, a non-magnetic mineral. These high-angle faults thus show up as slight lows aligned in a straight line."

In my early magnetometer work, I had a bias toward magnetic highs, which were sometimes caused by hot magma contacting reactive rocks, such as limestone, and forming magnetite. However, I've learned that successive intrusions into an area are generally more felsic and less magnetic as time goes on, and that altered granites may produce magnetic lows. Often identified on geophysical maps by their roughly circular-shaped areas of magnetic lows, these granites are the latest targets for gold exploration in Interior Alaska. Copper granites in southwestern United States have a "donut-shaped" magnetic signature because magnetite has been destroyed by alteration in their interior. While the outer edges show a high from magnetite created in the contact zone.

"Some lows may be buried plutons," Burns noted, and unfurled maps of the Manley Mining District. Like a circular South Sea island blue-green lagoon area, the tin granite of the district stood out as a magnetic low. Nearby an oval shape was only faintly green with more yellows and oranges, possibly a deeper intrusive body that didn't have as low a signature because of overlying strata.

Geophysical maps give the prospector another way to examine the ground. Relating the signals of known mineralized areas to more distant locales may suggest new places to explore. In a similar way, after the first diamond pipes were found in the Northwest Territories, other companies overflew the area to obtain data on the kimberlite pipes' magnetic signatures. With this data they could do airborne geophysical surveys in surrounding areas for other diamond pipes.

Not all areas have geophysical maps, and a prospector may want to rent or purchase geophysical equipment to do a ground survey. One supplier that rents such equipment is Terraplus. They have a catalog, CD, and website that lists a complete range of geophysical instruments, rental fees, and used equipment prices. For more information, write to:

TERRAPLUS
625 Valley Road
Littleton, Colorado 80124
(800) 553-0572 or (303) 799-4140 / (303) 799-4776 Fax
http://www.terraplus.com

USING A MAGNETOMETER

So much of Interior Alaska is covered by wind-blown loess, gravels, and moss that I often borrow a portable magnetometer to check for subsurface features during my early reconnaissance of an area. The magnetometer has a sensor on top of a pole with a wire connection to the meter. As I walk cross-country, I stop every 200 to 300 feet and take a reading by holding the sensor pole upright, and pressing a button on the meter. In seconds I get a number, say 56,809. A dial on the box is set at 56,000, the normal magnetic background for this area. I note the number 809, and walk to the next station. If the reading goes up to 57,100 or drops to 56,650, I do a grid over the area to obtain a sketch of the trend, shape, and extent of the anomaly. These are relative values, and on another day may vary, but the high or low readings should be verifiable.

I make a loop or return to a marked place or base station every so often to check for any magnetic changes in the original reading due to solar storms or daytime drift (diurnal variation). As I do a traverse, I'm looking for changes in the signal which may indicate a new rock unit, a fault, a dike, or intrusive. A steady set of similar signals tells me I'm in the same rock type. I usually work from known bedding or an intrusive into covered areas where the subsurface is obscured. Since magnetite is a minor component in sedimentary rocks, a magnetic survey in a sedimentary area only measures the responses from the deeper basement rocks, buried intrusives, or faults.

Ground magnetometers are also used to map gold placer channels when magnetite, or black sand is present. Stations twenty-five feet apart (or closer) are marked perpendicular to the trend of the valley. Each of the lines crossing the valley may be 200 to 400 feet apart, depending on the amount of detail wanted. Small to moderate signal spikes (ten to thirty gammas) will target the magnetite. However, if the bedrock is complex, readings will be erratic. In Interior Alaska where the ancient streambeds are often buried and there is a good amount of magnetite, a ground magnetometer survey may locate the gold-bearing channel. Drilling or trenching is then used to confirm the data and compute the pay-channel values.

PROFESSIONAL GEOPHYSICAL SURVEYS

I have also contracted with a professional company for a ground geophysical survey when I found an area with several indications of mineralization. Magnetic and electromagnetic (EM) surveys were done at the same time, giving two sets of data that eliminated some types of false targets. Before the team arrived to do the survey, I measured out and flagged a grid across the area. When readings were taken at each station, the flagging identified the location. They gave me print-outs of data from each traverse, and then discussed the interpretation with me.

A magnetic survey measures magnetic responses of the rocks, while the electromagnetic survey measures resistivity and conductivity. A water-filled fault or graphitic strata will be very conductive and give a high EM response, but the magnetic response will usually be flat, unless the graphite has magnetite. Massive sulfide deposits also are conductive, and some may have magnetite, so the magnetic and EM highs will frequently coincide. Copper porphyries have had their magnetite destroyed by hydrothermal fluids, and may show a magnetic low. They would also be more conductive due to disseminated copper or molybdenum and would have a high EM response. Having both measurements together allows more reliable geologic interpretation. In the geophysical survey I contracted out, the magnetic response was high, but the EM response weak. I later determined that the area was only schist with magnetite.

AIRBORNE MEASUREMENTS

Canada has many examples of mineral deposits that have been located using airborne and ground follow-up geophysical measurements. The Canadian Shield rocks cover almost half of that country and date from the Precambrian era, which represents eighty-five percent of geologic time, a vast period during which a variety of mineral deposits could have formed. Two-mile thick glaciers ground down and flattened much of the northern terrain and after they melted, many geological features were gradually buried by gravels, sand, windblown silt, clay, deep muskeg, and swamps. Before airplanes and airborne geophysical instruments, prospectors laboriously portaged between the lakes and streams, enduring black flies and mosquitoes to locate scarce outcrops in their search for mineralization.

In Canada in the late 1950s, airborne magnetic and electromagnetic surveys led to base metal discoveries such as those at the Mattagami Lake Mines in Quebec. More than two hundred and thirty conductors were identified with the Watson Lake group having both electromagnetic and magnetic responses. After follow-up ground surveys pinpointed the location of the anomalies, the anomalous sites were drilled. Under 50 to 100 feet of clay and gravel and 200 feet of barren volcanic rocks, the drillers found rich zinc and copper ore and associated gold and silver. The ore mineral chalcopyrite gave good electrical conductivity to the ore body, and magnetite produced a corresponding magnetic high.

Geophysicist Laurel Burns, Alaska Division of Geological and Geophysical Surveys, examines a map produced by an airborne electromagnetic survey. Subsurface features, such as faults, changes in rock units, and conductive base-metal deposits, can be identified.

My favorite geophysical story is told by Hugh Fraser in *A Journey North: The Great Thompson Nickel Discovery* (International Nickel Company of Canada, 1985). In the late 1940s and 1950s the International Nickel Company of Canada (Inco) was searching for new nickel deposits to replace the declining ones around Sudbury, Ontario, and they identified favorable rocks, peridotites, in northern Manitoba. Since these rocks gave high magnetic signatures, Inco could easily use aerial geophysical surveys to map major trends to follow up. Pyrrhotite, an iron sulfide, which often occurs with pentlandite, a nickel sulfide, created the high magnetism of the peridotites, and pyrrhotite is also a good electrical conductor, so a combined magnetic and electromagnetic survey pointed out areas or features with high potential for nickel.

Ground follow-up work was done in the winter when the lakes and muskeg were frozen. This was bitterly cold work, but the company persisted for seven years in the search. As each anomaly was located on the ground, camp was set up and drilling commenced. The company's advantage came from the use of airborne geophysical surveys, which others didn't have at the time. Inco could claim promising areas based on the geophysics.

In 1956, a low swampy area with black spruce around Cook Lake, about thirty miles northwest of the Canadian Railroad from La Pas to Churchill, was selected for drilling. Crews had been hauling equipment past this area on a winter road for years. They drilled into massive pyrrhotite with sparkling, golden "eyes" of pentlandite. Spectacular drill intercepts of six percent nickel were their reward after $31 million was spent on exploration. Nickel mineralization was outlined over a twenty-four-mile length, and three economic ore bodies were found by drilling. The town of Thompson Lake was built, as well as a $200 million mining, smelting, and refining complex.

Prospectors don't have the big exploration and drilling budgets of an Inco, but they can benefit from this example. The Thompson Nickel Discovery wasn't found on the most obvious geophysical target. It did not have a really good magnetic high over a conductor because the host rock was graphite rather than peridotite. Subtle anomalies don't always get the attention they deserve, and in this example, the company didn't get to the anomaly for seven years. Other prospectors were staking in the area, and had almost claimed the ore body before Inco got to it.

SUMMARY

Mining companies and geologists use a variety of geophysical tools to understand the geology of an area. Even with good outcrops and a well-exposed strata, these instruments can probe beneath the layers to give information about hidden features. For sediment-filled valleys, extensive alluvium on lower slopes, ground cover from deep weathering or vegetation, geophysical air and ground surveys, and maps reveal new features to explore. Sensitive electronic equipment that measures the physical properties of rocks is reaching deeper to find ore deposits. A prospector doesn't need to be a geophysicist to use these new instruments or to read the maps. They are like a crystal ball, a third eye that can see into the subsurface world.

23. Uranium: A Second Look

MOAB, UTAH URANIUM

Moab, Utah was transformed by the uranium rush in the southwestern United States during the 1950s. Now famous for its Colorado River rafting, canyon country, four-wheel drive adventures, and mountain bike trail "hikes," Moab seemed crowded when Dale Edwards came in the mid-1950s.

"Uranium did a lot for this town." Edwards' voice echoed in the deserted offices of the old Atlas Uranium Reduction Plant across the Colorado River from Moab. "The uranium industry made this town what it is today. When I came here during the uranium boom, there was no place to stay. I lived in a camper on the river for eighteen months until I got a place in town."

Yellow hard hats decorated a coat rack by the door. Through a smudged window the beveled layers of soil covering the old tailing pond blended with the distant canyon walls.

"There were about six hundred and fifty people in the mines and mill. Now there's just one—me. And when the Nuclear Regulatory Commission [NRC] trustee takes over, there'll be none. I came here and helped build the mill in 1955. I started as a laborer and cement finisher. When the mill started up in 1956, the Uranium Reduction Company asked me where I'd like to work and I said the laboratory. I was trained by the Atomic Energy Commission, (AEC) in Grand Junction, and I took correspondence courses. Eventually I became the radiation coordinator of the mill and mine."

Edwards showed me a long checklist he fills out for the NRC: air samples, radon gas, wipe tests, surface water samples, ground water monitoring wells, soil and foliage samples, and on and on. On his shirt he wore a badge that measured beta and gamma exposure, and on his belt was a box that took air samples.

Outside of Moab, Utah, a uranium tailings pond is a reminder of the uranium boom in the Four Corners and Colorado Plateau area of the southwestern United States. Once sought after by prospectors, uranium is no longer a popular target for exploration.

"When Uranium Reduction was first built," Edwards explained, "Charlie Steen owned some fifty percent of it. Charlie got some backing and built the mill. He was real generous with his money. He was a good man. In 1962, Atlas bought the plant from him and Uranium Reduction. At that time we had ore with copper in it. Our products were not only uranium, but vanadium and copper as well."

Edwards led me down an empty corridor past silent offices. "We shut down in March, 1984. We had a lot of ore, but we couldn't make a profit," he said as he had me lift each foot so he could check my soles with a scintillometer and alpha meter. "Just because we're down doesn't mean there isn't a lot of work to do. All of these checklist items are NRC license conditions. They have inspections at least once a year, and they can come in as many times as they want, unannounced. We haven't had any citations in many years. In 1988, the NRC gave us the go-ahead for decommissioning and capping the pond, but conservationists wanted an environmental impact statement. It came out this year–said there would not be a detrimental effect on the environment or human health."

He had me check a form verifying that I had been tested. "Still, a lot of environmentalists want the tailings pond moved. I can show you assays from our stations on the Colorado River above and below the mill, and there are higher readings above it. After a rainstorm, it's ten times higher uranium above the mill. There's a lot of uranium in this country. It outcrops everywhere."

URANIUM CRASH

After the big uranium prospecting rush of the 1950s, United States experienced Three Mile Island in 1979 and saw the movie "The China Syndrome" in the same year. Then we had the end of the Cold War, and on-going protests against new nuclear power stations. The price for uranium plummeted. Health risks of uranium mining, tailings disposal, and nuclear waste storage became issues. Ralph Nader founded the watchdog group Critical Mass and wrote *The Menace of Atomic Energy* (1977). An informal faculty group at Massachusetts Institute of Technology started the Union of Concerned Scientists in 1969 and published *The Risks of Nuclear Power Reactors* (1977), *Radioactive Waste* (1980), and *Safety Second, the NRC and America's Power Plants* (1987).

EARLY DEVELOPMENT OF THE SCINTILLOMETER

A half-mile from the Atlas plant offices and a block up Center Street in Moab, away from the anthill frenzy of sport utility vehicles and mountain bikers on Main Street, I found the Dan O'Laurie Canyon Country Museum. The uranium mineral carnotite, "red cake" from vanadium, and "yellow cake," containing about seventy-five percent uranium, are displayed, along with a Babbel counter, early Geiger counters, and scintillometers. The history of uranium mining is explained, and photographs capture the big operations: Standard's Big Buck Mine, Charlie Steen's Mi Vida Mine, and the Atlas/URECO Mill where uranium ore was turned into enriched "yellow cake."

At a Moab Points and Pebbles Club Gem and Mineral Show at the Old Spanish Trail Arena, club president June Cleveland filled me in on the Babbel counter, while she welcomed people to the event.

"My ex-brother-in-law, Gordon Babbel, read about Geiger counters in a national mechanics magazine." She jumped up to greet old friends and then continued. "He ordered one and when it came in the mail, it weighed 860 pounds! By the time he died, he was making one the size of a cigarette package. He also perfected the scintillometer. He mounted one on his airplane, and he could tell where there was uranium."

Prospectors use either a Geiger counter or the more sensitive scintillometer. However, neither of these instruments distinguishes between uranium, thorium, and naturally occurring radioactive potassium-40.

In addition, surface leaching often reduces the radioactive response, so pitting or drilling is needed to reach fresh material. Airborne radiometric surveys are useful, but ground follow-up must be done to determine if uranium caused the anomaly.

URANIUM PROSPECTING

Uranium was discarded in tailings during a vanadium and radium rush to the Four Corners area and the Colorado Plateau in the 1920s. Vanadium mills in Colorado and Utah were converted to extract uranium for the Manhattan Project in the early 1940s, and one thousand tons of additional ore from the Belgian Congo were brought out of storage on Staten Island for constructing the atomic bomb. Controlled nuclear fission was demonstrated in 1942, and Hiroshima and Nagasaki, Japan were destroyed a year later.

The Atomic Energy Commission (AEC) began a massive procurement program with guaranteed prices for ore and new discoveries. These price supports ended in the 1960s, however, when the AEC began buying only ore developed before 1958. By then, hundreds of deposits had been found, and the market was saturated with uranium.

At the Moab Gem Show, Jimmie Walker gave me his business card, with its rainbow and clouds, which advertised "The Finest Utah Redwood." Wearing a bola made from petrified dinosaur bone, and a belt buckle crafted by his son from moss agate and more bone, he looked every inch the prospector and rockhound.

"When this area got into uranium production," he told me, "I studied it and began staking claims and leasing them to companies." Now seventy-one years old, Walker knows these canyonlands. "I came up with a few theories. I had a little 1954 Jeep. Later on I got a four-wheel drive pickup that could carry more supplies. We'd have sleeping bags. Weather'd get real bad, we'd stay in the truck under a tarp. We would be moving around a lot." Lines on his tanned forehead bunched up when he got excited.

"My first claim, cost $2 to record it. Today, you got $140 per claim. They've destroyed the little guy." The wrinkles on his forehead built up into an accordion.

"What you'd do is work on a lease—get a retainer royalty. I just never got enough claims into production. They found some pretty good uranium on my property, but the prices went down. Then there was the Three Mile Island, new information about the Nevada testing, and the passage of the 1976 Federal Land Planning and Management Act... Little people can't survive."

He paused, his forehead cleared up and he returned to uranium prospecting. "You gotta understand the local geology. First, you look at the formations. The clay unit in the Morrison Formation is composed of volcanic ash. The theory is the uranium originally came from the volcanic ash. Microscopic obsidian runs higher in uranium. When oxygenated water got on it, it leached out different minerals and freed the uranium."

We stood by the cases where deep red chunks of his agatized redwood were displayed. "Our local geology is the Paradox Basin," Walker continued. "Two to three hundred million years ago it was an inland sea. Evaporation happened some twenty times, forming potash salts. When the sediments piled up, the salts flowed, affecting the surface, creating faults that made the Moab Valley, Salt Valley, Paradox Valley, and Lisbon Valley. There were lesser structures that didn't collapse. All of these helped develop river drainages. On the north side of the Paradox Basin, the Uncompaghre Mountains are still coming up. Pressure made anticlines that eventually fell in, exposing outcrops for prospecting.

"My prospecting was done visually, working on why or why not a stream system went where it did." He rubbed his forehead and smiled. "Now, Charlie Steen was an oil geologist. He saw the relation between uranium deposits and oil deposits. He felt you needed to prospect around oil structures, like in Lisbon Valley. There were already some small mines there on the surface. He drilled into the Mossback section of the Chinle Formation."

Charlie Steen, an unemployed geologist from Texas, had figured the area around Moab, Utah had potential. In 1952 he staked twelve claims near other discoveries, but well back from the usual target of the canyon's edge. His idea was based on the petroleum geology of anticlines as traps for fluids. He drilled into pitchblende, a dirty gray area in the sandstone and conglomerate. His shaft, put in on a shoestring and one drill target, found black ore running one-third of a percent to five percent uranium, after sixty-eight feet of tunneling.

"Charlie Steen packed them drill cores around in his pickup for days before he had a friend put a counter on it," Walker recalled. "You can't eyeball it, you need a counter to see the grade. A lot of people had a scintillometer because it was more sensitive. I had a Geiger counter and I used it for mining–to probe. When you were underground, uranium ores were oxidized, dark. The price was based on the higher grade."

Highly publicized, Steen's fortune was the kind of success story the AEC needed. Steen started the uranium processing mill near Moab, and became the largest employer in the county. In 1962 he sold his holdings for $10 million, but by 1969 he was bankrupt, because of his lavish spending, his generosity, litigation costs, and tax accounting problems with the IRS. His story is told by Raye Ringholz in *Uranium Frenzy* (1989).

NUCLEAR POWER

Nuclear power plants that produced steam to generate electricity (one pound of uranium generates the heat equivalent of about eight tons of coal) moved from the experimental stage to commercial operation in the 1960s. Alvin Weinberg, former director of the Oak Ridge National Laboratory wrote in his autobiography, *The First Nuclear Era* (1994), "We all wanted to believe our bomb-tainted technology really provided humankind with practical, cheap, and inexhaustible energy...." (Weinberg, p. 135).

As part of the optimism of the time, in 1974 the Atomic Energy Commission projected that fifty percent of the world's electricity would be from nuclear power by the year 2000. However, by the end of 1993 Canada's twenty-two reactors supplied only seventeen percent of that country's electricity, and in 1994 the United States had one hundred and nine nuclear power stations supplying only twenty-two percent of our electrical power.

Weinberg felt that the big four nuclear reactor vendors (General Electric, Westinghouse, Combustion Engineering, and Babcock and Wilcox) offered reactors at unrealistically low prices and built large reactors before lessons from smaller reactors could be incorporated into their new designs. Maintenance, expensive components, high construction costs, government regulations, and the unresolved issue of nuclear waste storage burdened the nuclear energy industry.

Developed countries without low-cost coal resources—France, United Kingdom, Japan, South Korea, and Taiwan—embraced nuclear power. According to Weinberg, these countries "kept the nuclear light burning brightly, while it gradually dimmed in the United States and other participatory societies" (Weinberg, p. 274).

URANIUM MINERALOGY

Uranium is widely dispersed in the natural environment, occurring on average from one to two parts per million in the earth's crust and four parts per million in granitic rocks. Of several uranium minerals, only thorium and uranium are found in quantities in nature. Uranium originally comes from magmas and igneous rocks, primarily granites. As the magmas crystallize, accessory minerals such as zircon, monazite, and apatite take up uranium. Late magmatic products such as pegmatites and veins contain uraninite in the form of pitchblende.

"I had some 'pitch' in the office once," Andy Harman, a Canadian mine maker recalled. "Just a handful from the Yukon Territory. I turned the scint [scintillometer] on and it just roared." Even outside the office on a window ledge, it sent the meter dancing, and he had to dispose of it.

Some uranium is deposited as film on rock crystals and these readily oxidize and release water-soluble uranium into solution in surface and ground waters where they adhere to clay minerals and on carbonaceous material. The resistant accessory minerals are washed into sedimentary beds, especially sandstones. Freshwater streams and stream channels that were interbedded with mudstone layers provided ideal beds for fine-grained uranium minerals, and deposits formed as uranium-rich minerals moved through the pore spaces in the host rock, replacing plant fossils and sand grains and cementing the minerals. Minerals that are commonly associated with uranium include copper, vanadium, selenium, molybdenum, lead, and manganese.

Granitic sources of the uranium were often remote from the sandstone and mudstone beds, and the method of deposition wasn't well understood for years. Flat-lying bodies of uranium were found parallel to the bedding, usually in just one stratigraphic layer. Broad shallow basins entrapped the uranium-rich water and sediments. Ground and stream water normally carry uranium to the ocean where it precipitates in muds containing organic material or in phosphates.

Uraninite, an oxide usually seen as pitchblende, and coffinite, a silicate, are the most common primary uranium minerals. Both are black, and occur either where they have replaced fossil wood and sandstone, or where they are disseminated in the sandstone pores. Pyrite or marcasite is found in nearly every unoxidized uranium deposit in sandstone. The reducing environment of the groundwater traps causes ferric oxide to change to ferrous oxide, which forms pyrite.

A variety of secondary minerals are created from the primary minerals and from migrating ground and surface waters. Most people recognize the yellow vanadates, carnotite, and tyuyamunite. Navajo and Ute Indians were able to lead geologists to some outcrops of red vanadium and yellow carnotite because they used the colorful minerals in ceremonies and artwork. There are many uranium arsenates, carbonates, phosphates, and silicates; thirty new uranium minerals were found in the 1950s.

Sandstone-hosted uranium deposits are related to the flow of mineralizing fluids through the host rock. The most common form is a "roll front,"–a single roll that is C-shaped in cross-section. Other deposit shapes that occur are pods, blankets, and logs. Mineable ore grades in the Four Corners area were from one-tenth to one percent uranium oxide. In contrast, ore from Canada's Athabasca Basin mines begins at over one percent uranium, and some new mines average fourteen percent uranium.

Key features of productive ground are thick sandstone beds; red sandstone turning to light shades of brown, yellow, gray, green or white; anomalous radioactivity; carbonized plant remains; colorful secondary minerals of iron, vanadium, copper, molybdenum, and cobalt; and a mix of sandstone and mudstone rather than clean sandstone.

Uranium deposits form in the thicker and more permeable parts of sandstone beds. A sandstone strata of one hundred to one thousand feet thickness might contain an ore-bearing layer less than one hundred feet thick, but sandstone less than forty feet thick generally is not favorable for large ore bodies. The many colors that are typical of a uranium deposit result from the reduction and oxidation of the secondary minerals in the sandstone and mudstone and may extend well beyond the deposit.

When I asked Lin Ottinger during a visit to his Moab Rock Shop, if he had any uraninite, he led me to a back corner where a few pieces of rock and organics lay in a small box.

"I had one fellow come in here with a counter," he recalled, tucking a black specimen under his belt. "I told him I ground the stuff up and ate it for my health." He chuckled like the seventy-year-old-kid he was. "He put the meter up on my chest and I pulled it down to my belt where I had the rock hidden." Ottinger grinned. "He claimed I broke his machine!" He pulled the rock out of his waistband, put it back in the box, and turned serious. "Some people think that stuff's good for you, but I don't believe it."

UNHEALTHY EFFECTS OF URANIUM

Approximately thirty percent of uranium mines are open pit. However, underground mines have significant health hazards because of radon and particles coated with uranium minerals. Colorado Plateau mines often hired Native Americans for underground work, and in the early years of mining, health dangers were not recognized or understood. Peter Eichstaedt exposed the long-term health problems that Native American miners are now having, in his book, *If You Poison Us* (1994). Steve Voynick's, *The Making of a Hardrock Miner* (1978), described the dangers of mining uranium in unconsolidated ground, and the sorry state of early inspections of mines for compliance with regulations.

Ralph Nader divided the use of nuclear power into the "front end of the fuel cycle"– (Nader, p. 82) mining, milling, and conversion into uranium fluoride gas and enrichment of uranium, and fabrication into fuel–and the "back end" Nader, p. 143) of the fuel cycle–the treatment of nuclear fuel once it's removed from the reactor. Nader pointed out the worst hazards of mining: that radon daughters (polonium 218 and 214, and bismuth 214) resulting from poor ventilation remain suspended in air or on dust particles and cause lung cancer, and that contaminants from tailings piles polluted the ground water, plants, and animals.

Such concerns about mine safety, environment, and health have led to stricter regulations and better monitoring. New mines now must carefully plan tailings disposal, underground mine ventilation, and mine reclamation. Because of the high uranium content in the ores of the Saskatchewan mines, Cogema Resources has developed special remote underground mining techniques: the area to be mined is frozen, and high pressure water is used to cut up the ore body and pump it out.

URANIUM MINES AND MINERS

To visit a uranium mining area, I drove thirty miles south of Moab, stopping briefly at the Wilson Arch to photograph this natural wonder by the roadside, and then farther on spotted a modest sign: "Lisbon Valley Industrial Park." A paved road led into picture-perfect canyonlands. Off one road were UNOCAL's helium, propane, and nitrogen fields. Union 76 had several more small wells tucked into the canyons. I went along the straight Charlie Steen Road for about six miles until it wound into a narrow streambed. The road got rougher as I ran into spring runoff gravel and sand. When I spotted a mine dump with a train of abandoned ore cars off to one side, I parked and began to walk up to the rim.

The favorable oil and gas structures in the area and Big Buck Mine up on the rimrock were clues that Charlie Steen followed in his search for a uranium trap. In a low saddle that was covered with sagebrush, I saw cleared areas and a caved-in mine entrance. Steen's Mi Vida Mine was an extraordinarily rich discovery. He brought his family up here while he

Uranium mines in the southwestern United States are largely inactive. Charlie Steen founded his Mi Vida Mine behind this canyon rim near Lisbon Valley, south of Moab, Utah.

drilled, and they shared the loneliness and deprivation of living so far from town. Now the area has been completely restored, and no sign of uranium tailings remains.

On a typical sunny day in Moab, I met Verle Green outside the trendy Moab Diner, and he told me about working in the uranium mines. "I started out in 1954 in a little mine in Colorado on Tenderfoot Mesa. I came over here to La Sal to work for the Hidden Splendor Mining Company, and I worked at the Ike Nixon Mine. They got more high grade ore out of that than any mine in the area," he recalled. Green wore old cowboy boots and a corduroy jacket, smoked Marlboro Lights, and had a hearing aid.

"I worked at all the mines. I quit in 1961. I drilled, blasted, and ran equipment underground. We didn't use ear plugs. We had respirators, but they'd plug up." He put out his cigarette and field dressed it down to nothing. "A mine is as safe as you make it," he pointed out. "Main thing when you went in—test the roof, bar down loose rock, and get rock bolts in."

He'd been a deputy sheriff in San Juan County, Utah, worked in the oil fields as a roughneck and roustabout, ran rafts on the Colorado River, guided four-wheel drive back-country tours, and worked in the Grand Canyon on the mule rides.

"I was what they called a contract miner. We had a guaranteed day's pay, but we could double our wages. I only spent seven to eight years in it. I've got friends, fellows I worked with, and a lot of them are gone. They all developed problems over the years. Many of them got silicosis, which turned into lung cancer and pulmonary fibrosis. When I quit in 1961, I had more seniority at Atlas Corporation than anyone on the payroll. Yeah, I loved it. I would've stayed but I got bronchitis from the diesel fumes and dust. It probably saved my life."

He looked at the busy street traffic. The restaurant was too full to find a table.

"Back in the 1950s, Moab was boomin'. Everybody had jobs, new cars. It grew overnight to about five thousand. After Charlie Steen made his big strike, there were promoters, claim stakers—everybody was out to get rich."

He shook his head. "All we got left is tourism. People comin' in here are buying up property. People are always gettin' lost in the canyonlands. They have more search and rescue in this county than anyplace in the United States. The mining companies are environmentally conscious, but uranium here is shut down."

URANIUM DISCOVERIES WORLDWIDE

Aside from the Colorado Plateau deposits, other important sources of uranium include veins, paleoplacers of South Africa, the strata-bound Olympic Dam deposit in Australia, the rich sedimentary deposits in Saskatchewan's Athabasca Basin, and the Elliot Lake-Blind River paleo-conglomerates of Ontario, Canada. In the early years of production, uranium came from nickel-cobalt-native silver veins in Czechoslovakia, the Belgian Congo (now Zaire), and Canada's Great Bear Lake. These hydrothermal veins contained uraninite (mainly in the form of pitchblende), davidite, brannerite, and coffinite.

South Africa's Witwatersrand, or Rand, has produced uranium as a by–product of mining gold from cemented conglomerates, where uraninite and gold are closely associated. In a similar type of deposit, Canada's Blind River-Elliot Lake Precambrian river channel conglomerates north of Lake Huron contain uranium, thorium, and rare-earths. Though the eleven mines exploiting this thirty-square-mile Canadian uranium field are now closed, the potential for other deposits remains high.

The 1975 discovery in South Australia of the Olympic Dam copper-uranium-gold-silver deposit has added enormous reserves to the world's uranium supply. Extending over fifteen square miles and with a vertical thickness of mineralization up to one thousand feet, this sediment-hosted deposit contains an estimated two billion metric tons of one and six-tenths percent copper and six hundredths percent uranium.

Canada leads the world in uranium production, and all of that comes from Saskatchewan. Discoveries in northern Saskatchewan's Athabasca Basin since 1968 make this one of the world's premier uranium environments. Radioactive boulders led geologists to uranium source rocks. The enormous McArthur River and Cigar Lake deposits average fourteen percent uranium. Formed in a sedimentary basin, the uranium was apparently leached from sandstone and granitic sources, transported through permeable beds, and precipitated in reducing environments in a process similar to the Colorado Plateau type of deposit. Cameco Corporation and Cogema Resources, a subsidiary of Paris-based Compagnie Generale des Matieres Nucleaires which have their head offices in Saskatoon, are the majority owners and operators of many of these mines.

SUMMARY

New discoveries and increased production in Canada have made that country a world leader in uranium mining and nuclear technology. Demand for uranium is rising. Even though there are no new nuclear power plants being built in the United States, most of the electricity in France and the United Kingdom is now produced using nuclear power. Is the world ending an experimental phase and entering a new nuclear era, a post-Chernobyl world in which uranium and nuclear power have come of age?

24. Developing a Prospect

"How much ya want for it?" a man in a suit and tie asked brothers Len and Dave Piggen over their display of rocks and prospect reports at the Vancouver Cordilleran Exploration Round-up. He swept their ultraviolet lamp over a tungsten-bearing rock and whistled at the blue streaks of scheelite. Crowds of well-dressed men edged past the table strewn with inviting chunks of mineralized rocks in the "Prospector's Tent," a room set aside for prospectors to display their findings and offer their claims. Len eagerly answered my questions, proud of their discoveries.

"We've got three properties: the Lucky Bear Claims, the North Claims, and the Spap/Head Claims," he told me. "Last year we sold the Cam/Gloria to Teck Corporation after we showcased it here. We had lots of people talking to us. We signed the final deal around May."

Len and Dave Piggen option their properties at the annual Vancouver, British Columbia Round-up. Mining companies are usually interested in property submittals.

INDIVIDUAL PROSPECTORS AND DEVELOPMENT

Looking beyond the initial discovery, a prospector has to either market a prospect or develop it, in order to benefit from the find. Unless the prospect is an easily mined vein, gold placer, or gem pegmatite, chances are a prospector will be looking for a larger company to pick it up. Even a small-scale deposit will require capital to evaluate its grade before undertaking a mining operation.

"For people who are interested, investors, we've got property details, management zones, maps, all the assays, color photos, and our biographies," Len told me as he pulled out a black binder and flipped through it. "This is the Lucky Bear property. We have expanded data, assays—the good, the bad, and the ugly—a claim summary, history." He stopped at a section of assays and pointed out the values.

"One of the claims had some work. We summarized the assessment reports and we listed past production from other mines in the area. With our limited knowledge, we get into a little of the geology, and use information from any site visits by other geologists and mining companies. Some of these guys had better assays than we did. Road access—we tell them how to get there from Kamloops, news articles, and other reports."

While Len showed me their detailed report, his brother Dave, fielded questions from interested executives. "The vein is hosted in the granite," he told one man. "This has a bit of gold in it and tungsten, too. We think the tungsten is a pathfinder for the gold."

Another visitor examined a rock with a hand lens, checked the magnetics, and asked for a property description.

"We haven't sold anything yet," Len confided. "I expect people to go back with the information we gave them and corporately make a decision. We've had lots of interest. That's what we want."

"The option payments [where a company pays a yearly fee to explore, with an option to purchase] really keep us going," Dave admitted.

Canada has two major mining conferences where prospectors can market their properties: the Vancouver Cordilleran Exploration Roundup, held in late January in Vancouver, British Columbia, and the Prospectors and Developers Association of Canada (PDAC) International Convention, in early March in Toronto, Ontario. Bringing properties to these conventions is a real test of a prospector's mettle. I attended the

Vancouver Roundup with a property write-up, but I didn't sit at a table with samples and work the crowd. I simply passed my report on to one of the many individuals connected to the venture capital world of the Vancouver Venture Exchange.

The conference costs about $105 (U.S. dollars), and I found an inexpensive hotel out East Hastings Street near Chinatown. The nearby affordable meals in Chinatown were a real treat: endless glasses of tea, noodle soups, chow mein, and a hundred other dishes. For information on future conferences, contact:

THE BRITISH COLUMBIA AND YUKON CHAMBER OF MINES
840 West Hastings Street
Vancouver, British Columbia V6C 1C8
(604) 681-5328 / (604) 681-2363 Fax
http://www.chamberofmines.bc.ca.

MARKETING A PROPERTY

Before a property can be marketed, all relevant information must be organized into a report showing that the property has been legally marked and claimed, adequate samples have been gathered, reputable assays have been obtained, and a reasonable-to-glaringly obvious mineral deposit of economic value appears to exist. Factors such as the property title being clouded, the partners disagreeing, access being difficult to impossible, the ground being underwater, or the mineral not being in demand, obviously place the property out of the reach of a prudent investor. I've seen a property submittal without any assays and with wild comparisons to deposits in other parts of the world; it gave the impression that no one had actually been to the site.

The location of the claim block, its topography, and the possible means of access answer the initial questions about where the prospect is and how difficult it will be to develop. Access is a major consideration, because of the costs involved in getting equipment in and ore out. In addition, an interested company would want to know who owns the land, which management agencies would require permits, what ecological and environmental constraints need to be addressed, what is the mining and exploration history in the area, what sites are currently being worked, and how is the property related to known deposits.

AN OPTION AGREEMENT

I staked my Old Dog lode claims on a ridge north of Fairbanks, Alaska because of a gold placer in the valley below. From an old USGS bulletin, I obtained cross sections of the placer creek and estimates of the historic gold channel's width, depth, and richness. A minor granite outcropping on the ridge had pyrites and a vein with anomalous gold and arsenic. An old antimony mine about half-mile away also indicated potential mineralization in the area, and a report on the antimony mine gave its production figures and discussed the local geology. In addition, my claims were between two known gold lode areas that had similar styles of mineralization.

For a few summers I mapped the granite exposures and did soil sampling along the hillside. The assays of my samples confirmed a large gold and arsenic halo. District geological studies gave me the regional geology, and supported the lode mineralization setting. I researched land ownership, a right-of-way to the property, and other nearby mining claims. From an overlook, I took color photos of the area and marked the location of the claims on the prints.

I knew a large mining company was evaluating the adjoining property, so I contacted them and arranged a meeting. I had my written report, assays, maps, and clearly established rights to the claims, but even though their geologist was very interested, the company didn't follow up after my presentation. After several letters to them, I realized I had to go elsewhere. It turned out that their upper management had decided to get out of gold exploration and had dropped their work in my area.

I eventually signed a "Mining Lease and Option to Purchase" agreement. This permitted a different company to "survey, evaluate, and drill for, develop, mine or otherwise extract, stockpile, store, process, treat, remove, ship, and market or otherwise dispose of any and all [s]ubject [m]inerals in, upon or under the [p]roperty...."

In such agreements, the initial payments remain low for the first few years (I received $7,500 each year for two years), and then they increase to $10,000 and top out at $15,000 if the company retains the claims for five years or longer. They also paid the annual claim rental fees, and I could use their expenses on my future annual labor payments. Within two years they had verified my soil anomaly, dug shallow trenches, drilled several holes to one hundred feet, and finished with six holes to five hundred feet.

All their data and drill cuttings were given back to me when they abandoned the property.

If the property had become a mine, I would have received a production royalty of four percent of the net smelter returns (NSR). If the company decided to buy the property, they had an "Option to Purchase" it for $2 million less any previous payments. The entire agreement amounted to twenty-five pages, and was written in an easily understood manner.

An agreement with a major company often involves a percentage of the net smelter return. Anywhere from two and a half to five percent NSR can be part of an agreement, and what is to be deducted from the NSR has to be spelled out. Companies will want to deduct freight costs to the smelter, taxes, and other charges before calculating the NSR. In my agreement, once they paid me $2 million, the property became theirs. Other agreements have provisions for selling the NSR, perhaps one to one and one half percent for $1 million with the remainder negotiable.

"We used to do a handshake agreement for one million in exploration," Andy Harman, a Canadian geologist recounted. "Now you need written agreements. I had an Australian friend. He sold a copper deposit, but he was too trusting and he made a sloppy agreement. A take-over of the company led to their ignoring his previous contract. His NSR was probably worth $15 million, but he could only get $2 million. The legal costs would have been too high, and he was seventy-one years old.

"A mid-sized company that is well-financed is the way to go," he believes. "The big companies will promise you the moon, drill your property like Swiss cheese, and then drop you in one year."

On the other hand, my experience with a big company having a commitment to my district was a positive one. I was informed whenever my claims were being drilled, and I often visited the site to discuss the findings with a geologist. They saved me tens of thousands of dollars in exploration costs, and made a realistic appraisal of the lode potential. I have all the data if I want to pursue further exploration or development.

Author Jane Gaffin interviewed Pete Risby, chief prospector with Welcome North Mines, in her book *Cashing In* (Nortech Services, Whitehorse, Yukon. 1980). "A cash payment now in the first phase would generally be from $3,000 to $10,000," he told her. "The greatest amount I ever received as front-end money for an individual property was $50,000, after its geological evaluation" (Gaffin, p. 218).

VENTURE CAPITAL

Another option is to investigate the venture capital markets represented in the Toronto, Vancouver and other stock exchanges. The markets have some "thieves" and "sharks," but if a prospector finds a good promotional group with a reliable track record and money behind them, there is the potential for a larger share of the pie than with a major mining company. This requires researching each company, their board of directors, their projects and financial statements, as well as how much time they spend in the field, and how much downtown.

"Lots of juniors have money in the bank," Harman noted. "Those that have the ability to finance a project and have money behind them are good to go with. Some groups have an excellent record."

"[T]his is a field exclusively for the experts," cautioned Robert Irwin in his book, *Profits from Penny Stocks* (Franklin Watts, New York. 1986). "It takes individuals with dozens of years of mineralogical and geological experience to be able to truly evaluate the potential of a company in this field" (Irwin, p. 161) He looked at each company's track record, its expertise, and its reserves: "Does the company have sufficient capital and expertise to start up production and become profitable?" (Irwin, p. 161).

"The venture capital market has discovered billions of dollars worth of mines," Harman asserted. He works with Vancouver companies and venture capitalists he knows, doing the fieldwork and feeding them the progress reports. However, it's important to keep in mind that venture capital follows the band wagon, working off new hotspots and fads, and that it dries up when the price of gold goes down.

Because of the expense of starting up a new publicly traded company, old "skeleton" companies may be refurbished with a new board of directors and new stock issued at five to one or ten to one to roll back outstanding shares. Then, according to Harman, You're "off to the races with a shiny new horse." A fast-paced account of the machinations and pitfalls of the Vancouver market was written by Matthew Hart in *Golden Giant, Hemlo and the Rush for Canada's Gold* (Douglas and McIntyre, 1985).

"The Vancouver Stock Exchange," Hart writes, "runs with cash. It is the great speculative exchange of the country, the place where the risk-takers congregate to try to sell one another their fancy pieces of paper. Occasionally, too, somebody slaps down real money and goes after a property with drills and a heart full of hope."

TIPS FOR SUCCESS

"The best thing," Harman advised me, "is to ask for a series of payments and a chunk of stock–it will have a one-year holding period before it can be sold. Depending on your costs, ask for $50,000 and $100,000 shares. They'll balk at the $50,000, but make it in two or three payments. The company will send their own engineer and he'll write a report and recommendations for an exploration program. Based on the report, the Vancouver Stock Exchange–now the Canadian Venture Exchange–will approve payment and stock. After one year you can sell the stock in small amounts, but you need a good broker when you're trading."

Sometimes it's worthwhile for an owner/prospector to spend some money drilling the property to get a better profile of what's there. Adding positive drill findings to a report can raise the value of the claims. Here, the costs are high, but the deposit justifies the expense. The drilling and sampling are done according to accepted practice and the results define the grade and extent of the ore. Junior companies often operate in this area, developing a good prospect with the expectation of a major company taking over.

"There's an order to things," advised driller Rocky MacDonald, owner of American Arctic Company in Fairbanks, Alaska. "When you have a raw prospect, do all the steps–[including] soil sampling and geological research. If you try to jump steps, you really up your risk level. It quickly gets beyond an individual, even a pretty rich individual. If I had a property, I'd go through soil sampling, even though it has considerable assay costs. That gets you a lot of information. Following the soil sampling, trench, or drill."

I generally dig in my heels at the expense of drilling. I prefer digging shallow pits with a pick and shovel, trenching with a backhoe or ditcher, or using a hand-held power auger because these methods yield good information at a lower cost.

"A person has to debate about that first hole–it may be a blank," Rocky cautioned. I caught him at a slack time in February when most of the drill rigs in his yard were still mantled with snow. He sat in an office cluttered with maps of Interior Alaska, and was constantly interrupted by the telephone. "I once drilled three holes up near Circle Hot Springs. They confirmed gold but didn't show a bonanza. The dollar return wasn't worth it. If I'd put in a dozen or fifteen holes, if I'd picked the right area, it might have made a difference."

Drilling is a major commitment of funds. On one project I spent $3,000 for a day of collecting cuttings from a reverse circulation drill rig. At the end of the day my arms ached from filling sample bags full of worthless schist. Drilling should be the last course of action, usually done after definite mineralization has been found.

DRILLING

"We're stuck in two dimensions to start with," Rocky said with a wry smile. "Drilling is expensive, but it really opens up the thing. Once you pry the box open when you poke that first hole, you see things you can't see on the surface."

The type of drill rig selected depends on the deposit and your funds. Track-mounted auger drills with a six- to eight-inch diameter screw can pull up soils, gravels, and soft rock from up to sixty feet deep. However, mixing and dilution of material occurs and the exact section where minerals were found will be only generally known. Plus, any water in the hole will wash cuttings off the screw before they reach the surface. I've used an auger drill to test a placer creek for the location of the gold channel, but I couldn't reliably compute the values in the gravels brought up.

"The Yukon guys have done auger successfully for placer channels," Rocky explained. "They have a situation that favors it—not many boulders, no groundwater on the benches, and a good sampling technique. You gotta pick the right tool for the job." He recommended the eight-inch solid auger to recover a sufficient quantity of gravels to compute placer gold values. I usually use a six-inch auger size because it goes down faster, and I can get more holes drilled in a day.

"In Interior Alaska, we also use auger drilling for penetrating the overburden to get a bedrock sample," Rocky noted. "The surface rock is covered so thoroughly here, not like in Nevada or Arizona where you have rocks sticking out. That will give you the top surface, but auger drilling is not suitable for going very far into the rock formation."

More expensive reverse circulation drilling, where the cuttings are forced up and out the drill hole by compressed air or water pressure, gives a fairly precise calculation of the gold values, and this drilling method is often used before actual mining begins.

"The two most common ways to drill are reverse circulation and diamond." Rocky went over the basic arguments for each. "Reverse circulation is faster and less precise, but you can look at more volume of rock. It can't do structural work–it can't look at fractures because the cuttings come up in fragments. Reverse circulation equipment is heavier than a diamond rig." These are big truck-mounted or track vehicles with a sturdy mast for the heavy twenty-foot lengths of drilling pipe, and with space for a large air compressor. As the drill bit cuts into the rock, air is blown either down the inside of the drill steel, or down the outside of the pipe, sending the rock fragments in a "reverse circulation" up to the surface into a collecting system.

"If you've got a remote place and steep terrain, then you'd use a [diamond] core drill. All the components are much lighter, but it's normally slower drilling. Geologists like to look at core." Drill core is a continuous record of the subsurface geology, and the one-inch or larger cylinders of rock are easy to examine for mineralization, changes in strata, faults, and alteration.

Rocky briefly spoke on the phone and then returned to the subject with an admonition from years of experience. "Any drilling program depends on the man on the machine. It doesn't matter what kind of machine. One guy will know how to get a perfect sample and the next won't." My own experience with drillers not familiar with permafrost ground bears this out. Where one man was forced to stop auger drilling because he caused the ground to melt and turn to glue, another knew the right speed and pressure to effortlessly penetrate the frozen soil.

"In any sampling system, you've gotta look at where the gold will get away from you," he warned. "There are all kinds of places that stuff wants to hide." For auger work, he drills through a hole in a sheet of plywood so the cuttings will be captured systematically and cleanly as they come out on the surface. Another common hazard was "salting down the hole," where cuttings from one level contaminated samples from deeper in the hole.

We exchanged stories about prospects we'd missed or let lapse. He, too, had spent time just across the hill from the Fort Knox Gold Deposit, and like me at that time, he was just learning the district.

"In the fall of 1987, my company drilled two holes for the owners," he remembered. "They were in ore the whole way. The next season we drilled 5,000 to 10,000 feet. That property traded up through several hands. Those first two holes showed they had a fish on. They spent around $17 million to drill it out and sold it to Amax for over $100 million. It was wealthy private individuals financing the thing. You look at the multiplier, they got ten to one."

SUMMARY

Between the prospector's find and the eventual mine are many branches along the path. Costs quickly escalate, but each step of the process will confirm or negate taking further action. Marketing a claim requires solid data, a good write-up, and an interested funding source. Taking it a step further, into drilling, may bring greater rewards, or a bust and big bills. "There are no economic rules attached to it," Rocky noted. "The prospector takes all the risk chasing a dream."

25. Interior Alaska's Gold-Rush: A Quiet Frenzy

Since the 1980s, the world-class Fort Knox and the True North Lode Gold Deposits were discovered near Fairbanks, Alaska, and now the incredible Pogo gold find in a remote area of Interior Alaska has been added. If this were an old time strike, the bell over the International Bar would be ringing off its perch. Jack London's hero Burning Daylight would be playing high-stakes poker and the Klondike Kings would be showing off their nuggets. The old days, when men had outfits at the ready and stampeded on the strength of a rumor, seem to be over. Is gold fever really dead?

Interior Alaska has a gold belt about 200 miles wide and 600 miles long extending from near Denali National Park to Dawson, Yukon Territory in Canada. The Great Tintina and Denali Faults bracket this area on the north and south. Historic gold districts include the Kantishna, Manley Hot Springs, Rampart, Livengood, Fairbanks, Circle, Richardson, Chicken, Eagle, and just across the Alaska-Canadian border, Forty Mile, Dawson, and the Klondike.

One hundred miles southeast of Fairbanks, a forty-eight-mile winter ice road branches off the Alaska Highway near the town of the Delta Junction, passing the summer cabins around Quartz Lake and snaking toward the newest and richest gold discovery—the Pogo. Rumors circulate about the exploration work being done by Teck Corporation and Sumitomo Corporation, such as, "Eighteen hundred feet wide and five thousand feet long and still open at both ends," "A rich layer cake of a deposit, richer than Fort Knox," "Probably the biggest find in North America," "Like Canada's Hemlo, or Kirkland Lake," and "Half-ounce to the ton!"

Andy Harman, a venture capitalist and mine-maker from Vancouver, thinks this will blow things wide open.

"People will be staking everywhere. Any prospect will be valuable. Get yourself a good claim block now," he advises. He's staking and promoting prospects in Interior Alaska as the Pogo excitement builds.

Evidence of the renewed interest in prospecting can be seen in many places. On a dirt road outside of Fairbanks, two men hastily leave a tan Suburban. One carries a power auger over his shoulder as they slip into the underbrush as quickly as possible. Everywhere, red fluorescent ribbons flutter on trees, leading off in straight lines across muskeg and through spruce forests. Occasionally, a helicopter passes overhead, with a torpedo-like "bird" hanging down on a cable beneath it.

In restaurants in the Alaska Highway towns of Delta Junction and Tok, men in hiking boots and khaki vests talk of GPS coordinates, drill rigs, and anomalies. The *Fairbanks Daily News-Miner* carries feature articles on new mining companies in town. Interior Alaska is in the middle of a gold rush, but where are the crowds and the high rollers on a binge? The biggest excitement is when a shift from the Fort Knox Gold Mine comes driving off Cleary Summit near Fairbanks and hits the Fox General Store for gas, a snack, and a six-pack.

A few Fairbanks prospectors, such as Rudy Vettor and Roger Burggraf, have done well over the years developing prospects they found. Never one to hide his talents for finding gold, Vettor bellows out his latest discovery to all within earshot.

"Veins shot through the granite! I got another mother lode!"

In contrast, Burggraf lets visitors handle some of his large gold nuggets, and generously gives away slices of quartz vein with visible gold. He speaks softly of his Fairbanks Mine and his work in the Brooks Range.

"We got new backers and they're doing a good evaluation," he says with a smile. "I think we'll see it in production soon."

THE ROLE OF THE PROSPECTOR

The new style of gold rush still begins with the prospector tramping over the ground, searching for signs of alteration or veining. It's a solitary occupation, with no one else around to share the camp or pass the time. During fall hunting season things can get dangerous as the hills fill with weekend warriors out for a quick moose or caribou kill. Then winter snows put an end to the quest, and other pursuits are taken up.

With luck, some targets emerge out of the prospector's sampling and assaying. A report with maps, data, and geology is crafted to present to a mining company. Since only a rich vein could be worked by an individual, most prospects are passed on to a well-heeled company for more testing and drilling before the expense of a mine is undertaken.

"We're interested in your Monte Carlo Prospect," a voice on the telephone says. "Let's meet to discuss a contract."

In an office the deal is signed. $7,000 to $8,000 annually for the first few years of exploration, topping out at annual payments of $15,000 after five years.

"We'll pay up to $2 million."

The prospector shakes hands, signs, and hopes it's a big one.

Two years later the company pulls out after drilling the hillside and not finding much besides arsenic. For a brief time the prospector dreamed of a fat bank account. Take the wife to Paris and see the sights. Now, he's back in the hills patiently cracking rocks and testing the soil. It's said that out of a thousand prospects, only one becomes a mine.

INDIVIDUAL PROSPECTORS VS. CORPORATE INVESTMENTS

What about the furtive pair with the power auger ducking into the roadside bushes? For $15 an hour they course over the hills gathering thousands of soil samples that will be assayed to generate a map of the area for their company. Crews working on such a regional and local reconnaissance program can cover large areas quickly, testing the soil, getting rocks from good looking outcrops, and panning stream sediments for concentrates. This is prospecting without passion, covering the ground to gather data for a computer-generated map.

In contrast, the independent prospector goes for the quick and dirty, seeking the ripest plums to test, and going where fancy leads. The corporate method requires a big budget and logistics. The solitary prospector has two good legs and time.

Professional claim stakers are called in to block out miles of claims. With compass or Global Positioning System (GPS) and bright flagging, they move in straight lines through dense spruce thickets and fire-tangled blow-downs. Some companies simply assemble the corner stakes with claim forms inside and have helicopters deliver the weighted bundle to each claim corner using a Global Positioning System. Having a "position" means having control. You don't have to buy out your neighbor if your claims already extend a few miles in every direction.

Around Fairbanks in 1998, there wasn't any land open to claim. Newmont Mining Corporation had a block. Barrick Gold Company bought into another. Placer Dome Incorporated controlled Ester Dome. Kinross Gold Corporation had the Fort Knox area.

In the old gold rushes miners' law specified only one claim per person per creek, except for the discoverer who got two claims. Alfred Brooks, Alaska's father of USGS work, pointed out the abuses of professional claim stakers tying up large areas using power of attorney during the Nome Gold Rush. In 1900 Nome, and continuing today, a discovery was supposedly required on every claim a person stakes. However, that rule was never well enforced in the past, and things haven't changed much, because large claim blocks are still the norm.

Outside of Fairbanks a helicopter maintains a fixed elevation above the ground, flying a straight-line coordinate while dangling a "torpedo" below. A mining company has contracted for a geophysical survey to map the magnetic and electromagnetic responses of the rocks. Altered granite will show up as magnetic lows because hydrothermal fluids destroy magnetite; whereas conductive sulfide beds of copper will make electro-magnetic peaks. Such expensive high tech prospecting is usually out of a prospector's reach.

In the Canadian Provinces, around Hudson Bay and the Northwest Territories, glaciation has flattened many features and obscured the bedrock with a covering of glacial till and permafrost muck they fondly call "loon dung." There, airborne geophysics has been used since the 1960s to locate many rich deposits. Similarly, in Interior Alaska the state is doing airborne geophysical surveys of high-interest areas using magnetics and electromagnetics to penetrate the mantle of moss and wind-blown Pleistocene silt and reveal the hidden bedrock geology.

MAJOR AND MINOR PLAYERS

If a prospector knows how to interpret the highs and lows and circles and stars on the maps, these geophysical targets can be followed up on the ground. Many turn out to be false leads like old bottle tops that set off a metal detector, and checking them out can be expensive. I once spent a summer drilling an anomalous magnetic high only to find it was made by a magnetic mineral that had no value.

Two hundred miles south of Fairbanks, near the Tok Airfield, a doublewide trailer houses a Kennecott Exploration Company field exploration office. From behind the building comes a mechanical thump and crack as a lab tech splits diamond drill core from a prospect in the Alaska Range. On a chalkboard inside the door of the office is written: "Some days you win; some days you lose. And some days it rains."

Andy Harman, in Alaska with the first wave of claim stakers, talks with the geologists about borrowing their helicopter. They gather around this stocky bear of a man whose fame is deserved.

"Yah, I was in on the Moosehorn discovery," he replies to one query. "Veins were sticking out all over the mountain, and there was eluvial gold on the hillside. Pretty rich, eh?" Changing the subject, he points to a spot on the map. "I need to look over this area near the Ladue River. It will take about an hour of air time to get in and return."

The Kennecott geologists scrutinize the blank area on the map, feigning disinterest.

"We're working just north of there," a tall, shaven-headed geologist admits. "We can spare the chopper." He measures the distance. "Our Tarus Prospect is only about twenty miles east."

"Copper-gold, eh?" Harman casually exposes their secret. There isn't much he doesn't know, from British Columbia through the Yukon Territory to Alaska. He spends his winters in Chile searching out prospects in the Atacama Desert among Indian farmers and blood-sucking Benchuga beetles, the carrier of Chagas' disease. He's the advance front of Vancouver interest, an indicator of where the next big "play" will take place.

That evening, after dropping him off, the geologists meet.

"What've we got on that area?" the lead geologist demands. "What's he know that we don't?"

"Let's do a recon after he leaves," the shaven-head geologist suggests.

In a remote valley, Harman and his leg man line out a row of flagging and square off trees as corner posts. A black bear briefly challenges these intruders, but one warning shot sends him packing. Clambering over blow-downs from an old fire they carefully mark out claim after claim. Even the moose trails are hard to follow in this tangled-up wilderness.

Harman spends a day on the creeks panning the gravels to obtain concentrates for assaying. Soon another blank spot on the map will be a claim block, an object of intense interest to those who follow this latest gold rush.

A year ago a converted log cabin home on the west flank of Ester Dome near Fairbanks served as an office for Placer Dome. Like horses tied to a hitching rail, a fleet of red four-wheelers was chained to the birch trees. After a day in the field, mud-spattered geologists headed for the beverage-laden refrigerator in the basement. A fresh pile of bulging sample bags on the porch showed off the days' booty.

"That's a dangerous slope on the north side," a lanky and long-haired field worker commented. "But the granite sticks out like a thumb."

"You working off Antimony Ridge?"

"Yeah. Down Nugget Creek." He broke into song. "Up on Cripple Creek she sends me... ."

Jeff Rogers, their lead geologist wore faded Carhartts and a crash helmet for driving "Tolovana," a four-wheeler with bent handlebars and a crooked frame. He explained how their $1 million budget paid for eight employees and several contractors.

"This is cookbook exploration. Most of the effort goes into soil sampling and follow-up drilling. Like a lot of others," he admitted, "we're wrapped up in the hysteria of gold exploration in Fairbanks." Scarcely a year later in the fast-paced world of up-and-down gold prices and corporate decisions, Placer Dome was gone having sold out their interest in Ester Dome to Kinross Gold–making Kinross the principal mining company in the area.

In a quiet Fairbanks subdivision with college street names, a weary field crew from the exploration arm of Kinross lays out their muddy sample bags in neat rows. The power auger is placed in a corner of the spacious garage.

"Enough rain for you today, Geaser?" the geologist asks a wizened figure in filthy jeans and t-shirt.

"We're swimmin' in it this summer for sure," Geaser answers. "And them rocks in the holes are tearin' up me back."

He runs the power auger day after day, pulling up soil samples to assay. If seen in the woods unexpectedly, he would qualify as a wild man, with his scraggly beard and stringy hair. A thin PigPen, Geaser provides the muscle and stamina to get the job done.

"After this prospect, we'll head on to Fox Gulch," the geologist tells him. "Then there's Caribou Creek to check out, and then the Richardson prospect." He jokingly asks, "What's that about your back?"

His wife, the modern version of a camp cook, prepares a hearty meal for the guys–juicy pot roasts, fresh breads, coffee and a rich dessert to keep the men energized. As summer passes, her husband slims down, becoming gaunt by the end of the season. The daily grind wears on them, and the living room refuge with its vaulted ceiling and big screen television is seldom used. Instead, a downstairs map room, with its computers, fax machine, and copier beckons. There are points to plot, reports to write, paperwork to do. It's a race through the brief summer before snow and frozen ground stop the probing and collecting.

A prospector comes by after dinner with samples and a map to go over. No one with a new prospect is turned away.

"I got these from where the fault intersects the dike," the bearded prospector explains. "And these schists are loaded with sulfides."

The geologist takes a loupe from his pocket and eyes the rocks. "We'll send them in for assay with the batch going out tomorrow," he promises. "We'll have results in about a week."

The big companies have their own fast track for assays; the individual prospector might get results in a month with luck. If the company pays for prospector's assays, they get the results faster and are the first to know if there's gold in the samples. So everyone benefits.

"Did you ever see that gold in the granite from the Democrat Lode?" the geologist asks. He pulls out a frothy, iron-stained rock with bright gold. "Look at those flakes!"

The prospector marvels at the generosity and whims of nature, wishing he had such a rock on his claims.

A year ago Newmont Mining Corporation had an exploration office near Fairbanks, on the gravel tailings from Gold Dredge Number 8. A small stone sculpture of a prospector panning above his sluice box and flowers in a massive dredge bucket greeted visitors. The company had four geologists out beating the brush around the seventeen square miles of claims that included the True North Deposit.

"We've assayed ten thousand samples over the last three years," reported Richard Harris, project manager. "We have a large claim block with a nice resource. I'm confident about Alaska's exploration potential."

Yet a year later they too sold out to Kinross and abandoned the district. Gold rushes are fickle: companies come and go. The bandwagon changes leaders, but the claim staking and searching continue.

In several papers about the area, Professor Rainer Newberry, a geology professor with the University of Alaska Fairbanks, called this the "North Star Gold Belt." In a recent Geological Society of Alaska meeting, it was called the Tintina Gold Belt. Whatever its name, the enormous potential of the region has been confirmed with the Fort Knox and Pogo discoveries. This may not be a return to the rowdy bars and Soapy Smith days, but the gold rush is in progress–just in time for the centennial celebration of when it all began in the Canadian Klondike and Interior Alaska.

CONCLUSION

At the end of an hour-long presentation about gold deposits to a high school class, an eager hand shot up.

"How do you pan for gold?" a bright young girl asked.

With a sinking feeling, I realized my audience wanted some practical methods beyond my models and theories, but teaching field techniques in front of a class or through a book is not the same as squatting by a creek with a gold pan. For a classroom presentation I could at least give a demonstration and bring in a few rocks to pass around. In a book, it's impossible to show how to tip a gold pan to wash off the top layer of lighter minerals, and these chapters can't duplicate holding an altered rock or tracing a vein.

So, I've concentrated on the steps to follow to find mineral deposits, and I've suggested additional resources to utilize. First, research an area to identify likely targets and land that is open for mineral staking. Then do careful fieldwork and sampling, and have your samples assayed. Even the most basic recreational gold panning can lead to lode discoveries if a person closely examines the heavies and semi-heavies in their pan concentrates. If an economic mineral shows up in the assays, mark and record mining claims, and follow up with more intensive sampling.

Each stage requires analyzing the data and making hard decisions about whether to spend more time and money. Gold fever is overshadowed by practical considerations. Luck and risk, fickle natural processes that don't always produce what we expect, and changing mineral values in the market are big factors.

A mineral discovery is more likely to happen when a prospector understands all the possible indicators of mineralization, including colors from oxidized sulfides, alteration in the rocks, and "kindly" geologic conditions. Prospecting techniques range from licking rocks to check for alteration to imagining the role of plate tectonics in the formation of an ore body. A successful prospector understands the potential of a limestone bed abutting a granitic outcrop. The ability to follow up the clues and a healthy curiosity are also essential in this process.

I frequently wander through the colorful open workings around Virginia City's Comstock Lode south of Reno, Nevada where the rocks illustrate clay alteration and hydrothermal veining. As I break rocks, my mind reviews the conditions contributing to the formation of these silver and gold deposits. Sometimes prospecting a worked-out area is as much fun as finding a new bonanza.

In the same way, driving past the roadcuts north of Fairbanks, Alaska, I never tire of seeing shiny muscovite mica glittering in the metamorphic rocks, and reconstructing the undersea basalts and beds of carbonate shells that now make up these greenstones and limestones. As I approach the placer creek headwaters farther from town, I notice a change in the metamorphic rocks from dark green oceanic rocks to brown biotite schists from continental sediments. On the mountains above the creeks I note the resistant granodiorite and quartz monzonite bodies, the probable source rocks creating the gold lodes and creek placers.

Understanding physical and economic geology and the unique factors behind the formation of an ore deposit adds to my enjoyment of the outdoors. I learned much of what I know about rocks from classes and by examining my local geology. Although in this book I often use Fairbanks and Interior Alaska as examples, that is only because it is convenient and familiar to me. Every locality has its own unique geology to study.

This introduction to geological processes and mineral deposits only touches on their amazing variety in nature. Attending prospecting and geology classes, searching out references, joining rockhound clubs, and subscribing to magazines are other ways to learn the subject. Taking field trips to investigate geological setting helps connect the names with the rock types and brings together people with similar interests. I once watched two busloads of geologists cheerfully go out into a soaking rain just to examine a new rock outcrop. Looking at the different compositions of sedimentary, metamorphic, and igneous rocks can be a lifetime study.

When I first encountered books on economic geology, I read and reread them because of the unfamiliar vocabulary. I bought a dictionary of geology, and gradually learned about plate tectonic environments, mineral deposit models, rock alteration styles, and the role of hydrothermal fluids. I've introduced some of this vocabulary in these chapters in hopes that my readers will find the related literature more understandable.

Each chapter listed relevant geology texts, personal accounts, reports, and sources. For example, Marshall Sprague's account of the Cripple Creek, Colorado Gold Discovery in *Money Mountain* relates practical information and is a true story. Robert Boyle's "The Geochemistry of Gold and Its Deposits" (Canada Geological Survey *Bulletin 280*) brings together a broad range of information about gold. I am constantly checking references and tracking down reports to learn more about a particular locality or deposit.

Once thought to be "black box" technology that is available only to trained company geophysicists, geophysical instruments such as metal detectors, Geiger counters (now scintillometers), and ultraviolet lamps are now commonly employed and have become important prospecting tools for everyone. I use a magnetometer in deeply weathered bedrock and soil-covered areas to trace potential faults, locate changes in rock types, and map highly magnetic deposits and strata. Such sensitive instruments, and magnetic and electromagnetic maps provide new information about the rocks and subsurface geology of an area. Of course, the rock hammer and hand lens continue to be essential too.

Mining companies employ the same relatively simple exploration procedures that individual prospectors use—following up geological clues and research to narrow down the targets; collecting and geochemically analyzing pan concentrates, stream sediments, soils, and rocks; getting consistent samples; keeping a record of what was collected where, with clear descriptions; assaying a variety of samples; and evaluating the assay reports and other information to determine potential values. Because prospectors often don't have big budgets, assaying many samples can be quite expensive, and clues should be carefully evaluated to narrow the assays to the most useful samples.

Companies welcome submittals from prospectors, but whether a company will be interested depends on their current corporate direction, their interest in the area, and other factors. A good property needs a solid report in order to sell it. The assays have to be high, the area must be correctly claimed, and its accessibility must be known. Nowadays, property option agreements are straightforward legal contracts.

The possibility of making a discovery drives a prospector to explore new areas. And ultimately, prospecting becomes a path of self-discovery—a test of your patience and resilience. It takes mental strength to walk away from a tantalizing find once it's been explored. Some days I'm elated by the certainty of success, and the next I'm kicking myself for not being realistic. Enjoy the quest, take the risks, and keep dreaming of the big one!

GLOSSARY

A

"A" Horizon – The uppermost part of a soil profile, usually composed of organic material, partially leached by water percolating downward.

Adit – An almost horizontal passageway into a mine.

Adsorption – Adhering to a surface. In stream sediments, metal ions are adsorbed onto the fine particles of silt and sand.

Airborne Geophysical Survey – Magnetic, electromagnetic, radiometric, or gravity survey done from an airplane or helicopter.

Alluvial Deposit – Sediment (gravel, sand, and mud) deposited by fresh water in a channel, stream delta, or coast.

Alteration Halo – Envelope or effect formed in the wall rock around a vein or ore deposit because of hydrothermal fluids, heat, or gases.

Andesite – Volcanic rock, the extrusive equivalent of diorite. Typical of a subduction zone environment.

Ankerite – Carbonate mineral composed of calcium, iron, and magnesium formed because of metamorphic pressure that produces blocky quartz with chlorite and possibly gold.

Anomalous, Anomaly – Above or below normal. A deviation from the average or background values.

Anticline – A fold in sedimentary or metamorphic strata with older rocks in the core. Strata slope downward on both sides of a common crest.

Apical – At the apex or highest point.

Aqua Regia – Kingly water: it dissolves the "noble metals," gold and platinum. It is also a mixture of nitric and hydrochloric acids.

Argillic alteration – Rock changed to clay within or near a mineral deposit. Sometimes subdivided into intermediate and advanced argillic.

Atomic absorption spectroscopy (AAS or AA) – Chemical analysis used in assaying. A solution of a sample is atomized in a flame, and the amounts of each element wavelength present is measured to obtain element concentrations.

B

"B" Horizon – In a soil profile the zone of accumulation below the A horizon, usually enriched in clays and aluminum oxides. This horizon may gain material by upward mobilization and percolation of soluble material from the underlying "C" horizon.

Back Arc Basin – Basin formed between an island arc and a continent where oceanic plate subduction underthrusts the island arc and continent, causing andesitic volcanism. Examples are Japan and the Philippines.

Basalt – Dark-colored, fine-grained volcanic rock. The extrusive equivalent of peridotite and gabbro. Hawaiian lava is an example of basalt.

Base Metals – In a classification of metals, the group that includes copper, lead, zinc, and tin.

Basin and Range (Horst and Graben) – Extensional area of continental rifting that creates elongated fault blocks forming uplifted mountains (ranges or horsts) and downdropped valleys (basins or grabens). Nevada's mountain ranges and basins are oriented north-south and extension in the region is east and west.

Batholith – A large body of intrusive igneous rock exposed over an area of at least 100 square kilometers (38.61 square miles).

Bench Testing – Trying responses (e.g. of a metal detector to different metallic minerals at home) before going out in the field.

Besshi-type Deposit – Japanese term for volcanic-associated massive sulfide (VMS) deposit, usually containing copper-zinc ore without gold or silver, in a back arc basin.

Black Smoker – Plume of hydrothermal fluids with black sulfide particles coming from a vent in an underwater oceanic ridge. The origin of some massive sulfide deposits.

Blind Lode – A deposit that does not reach the surface.

Blind Ore Bodies – Not showing on the surface.

Bonanza – A rich mine, vein, or ore body.

Bowen's Reaction Series – Explanation of igneous differentiation developed by N. L. Bowen. During crystallization of a magma, a sequence of minerals forms during the cooling process. For example, calcium plagioclase forms early, while quartz is late.

Breccia – Mixture of angular fragments in a matrix of finer particles (e.g., fault breccia, volcanic breccia, or kimberlite pipe breccia).

Browse – Leaves, twigs, and young shoots of trees and shrubs which animals feed on.

Bushveldt Complex – Very large layered igneous body in South Africa containing vast reserves of chromium, platinum group elements, and iron.

C

"C" Horizon – Soil profile zone consisting of partly decomposed bedrock and parent material from the overlying "A" and "B" horizons, grading into unweathered bedrock.

Calafite – Calcium-rich rims (especially on olivine) created during explosive transport of material in diamondiferous diatremes.

Caldera – A large depression formed by subsidence or collapse of a magma chamber, sometimes after a volcanic eruption.

Carlin-type Deposit – Disseminated gold deposit in carbonate rocks. Formed by hydrothermal fluids leaching minerals from underlying rocks. Named after Carlin, Nevada.

CAT – A brand of bulldozers.

Cheechako – A slang term for a newcomer.

Coarse Rejects – Parts of an assay sample spilt not processed or used by an assay lab.

Colorimetry – The analysis or measurement of elements or minerals by determining the intensity and hue of a color, as of a solution in chemical analysis, by comparing it with standard ones. See Dithizone.

Conglomerate – Coarse-grained sedimentary rock composed of rounded pebbles, cobbles, or boulders in a finer-grained matrix of sand or mud.

Contact Metamorphic – See *skarn* and *hornfels*. Metamorphosed country rock at or near an igneous contact.

Country Rock – Sedimentary or metamorphic rock into which magma or mineralization is emplaced.

Craton – Stable interior of a continental plate.

Crust – Outermost solid layer of the earth. Oceanic and continental crust have different compositions.

Cupel – A small, shallow, porous cup used in assaying gold, silver, etc, or an hearth for refining metals.

Cupola – A small protuberance from a larger igneous body. Cupolas may contain concentrated amounts of minerals.

Cyclone – A processing device used in mining and milling to swirl particles in liquid, to separate out higher and lower specific gravity minerals.

D

Diatreme – Vertical igneous intrusion in the form of a pipe, made up of breccia of country rock, magma, and mantle material. For example, a kimberlite pipe.

Diffusion Aureolas – The area around, alongside, or above a vein or mineral deposit containing alteration or more mobile minerals such as arsenic.

Dike (Dyke) – A minor igneous intrusion, often near-vertical and flattened, that cuts across the strata of surrounding rocks.

Diorite – Medium- to coarse-grained intrusive igneous rock with more than twenty percent quartz and containing plagioclase feldspar. Found in island arc settings.

Disseminated Deposit – Minerals dispersed throughout a host rock. A low-grade deposit such as a porphyry copper or Carlin-type gold deposit.

Dithizone – A chemical used in prospecting that changes color when mixed with mineralized soil. A colorimetic technique for locating mineralization.

Dun – A dull grayish-brown color.

Dunite (Peridotite) – An ultramafic rock composed of ninety percent or more olivine, often of mantle origin.

E

Economic Geology – Study of the origins, settings, and uses of minerals.

Electromagnetic Methods – In geophysical exploration, ground or airborne technology that broadcasts and receives an alternating current to map the electrical conductivity of the uppermost rocks in the earth's crust in the search for metallic mineralization.

Eluvial Deposit – Placer deposit formed from downslope creep of residual material over a deposit.

Epithermal Deposit – A mineral deposit formed at low temperatures (122°F to 392°F or 50°C to 200°C) near the earth's surface.

Excavator – A person or thing that makes a hole or cavity by digging or hollowing out. A dredge or steam shovel are examples.

Extrusive – Igneous body emplaced on the surface. For example, basalt, rhyolite, or andesite.

F

Face – Mining term for exposed rock surface being mined. Not the ceiling (back) or the floor. In an open pit, it is the pit wall.

Facies – Characteristic of a rock body that differentiates it from others.

Feldspathization – Sodium feldspars formed by high temperature alteration of rocks by hydrothermal fluids.

Felsic – Light-colored, fine-grained igneous rock composed of feldspar and silica. For example, rhyolite.

Femic – Ferromagnesian. Containing iron and magnesium.

Ferro-, Ferric – Iron-bearing.

Fertile Granite – Granite with high amounts of fluorine and lithium that has a high potential for forming rare-element pegmatites.

Float – Miners term for rock (usually mineralized) that has broken off from its source and moved downhill. "Tracing float" is the process of following alluvial and eluvial rock back to its source.

Flood Gold – Gold that floats and does not form placer deposits. Flour gold. The finest-grained placer gold.

Foliation – Compositional layering. Schistosity or layered structure in metamorphic rocks.

Fossil Placer – Ancient placer that has turned to stone. See *Paleoplacer*.

Fumaroles – A vent in a volcanic area from which smoke and gasses arise.

G

Gabbro – Coarse-grained igneous rock composed of plagioclase, pyroxene, and accessory olivine. Usually dark-colored. Mafic, from mantle-source rocks.

Geochemical Anomaly – Abnormal concentration of elements compared to a normal, or background, level.

Geyser – A fountain of hot water and steam that is periodically ejected from a surface vent. Usually found in areas where groundwater is heated by shallow magma.

Glory Hole – Large open pit from which ore has been, or is being, mined.

Gossan – A mass of limonite and other material remaining after sulfide deposits have been oxidized and leached away by percolating surface waters.

Granite – Coarse-grained igneous rock composed of quartz, feldspar, and plagioclase.

Granite Porphyry – Igneous rock with phenocrysts, usually feldspar and/or quartz in a medium- to fined-grained matrix.

Granodiorite – Coarse-grained igneous rock composed of quartz and feldspar in which plagioclase is more than sixty-seven percent of the total feldspar. Transitional between granite and diorite.

Graphic Granite – An intergrowth or sandwiching of quartz within large feldspar crystals in some pegmatites.

Greenstone – Dark-green metamorphosed mafic to ultramafic igneous rocks such as basalts.

Greisen – Masses of quartz and white mica formed by alteration of granite by hydrothermal solutions. May contain tin and/or tungsten.

Grizzly – A large mesh screen or other device with restricted openings, used to sort oversize rocks from material that is to be sieved for soil sampling, sent to a rock crusher, panned in a gold pan, or processed in a placer mining operation.

H

Heavy Mineral – Mineral with a specific gravity greater than 2.85.

Hip Chain – A small device attached to a person's belt that uses a spool of thin string and a counter to measure distances when walking.

Hoodoo – A natural rock formation of fantastic shape.

Hornfels – Dense recrystallized rock formed by thermal metamorphism at the igneous contact with non-reactive country rock.

Hydromuscovite – A white mica also known as sericite that is an alteration product in many mineral deposits.

Hydrothermal Alteration – Changes in country rock produced by fluids alongside veins or surrounding ore bodies. May extend from a few inches to a mile or more.

Hydrothermal Deposit – Deposit formed by precipitation from a hydrothermal solution. Not always an economic deposit.

Hydrothermal Solution – A hot, watery solution containing dissolved salts or other chemicals. Responsible for many types of mineral deposits. Some sources of the water include surface water, groundwater, sea water, and magmatic water.

Hypabyssal Rock – Igneous rock that crystallized near the earth's surface.

Hypothermal Deposit – Formed at high temperatures (572°F to 1112°F or 300°C to 600°C) and at depths of two to nine miles (three to fifteen kilometers).

I

Igneous Body – A volume of igneous rock within the surrounding country rock.

Igneous Rock – Rock that has solidified from molten material.

Immiscible – Unable to mix to form a single liquid.

Indicator Element – In a geochemical survey it is an economically viable component of the ore being sought. May be immobile or difficult to analyze, so the more widely dispersed pathfinder element is first sought. (See *Pathfinder Element*)

Intrusive – Igneous body emplaced at depth.

Island Arc – Islands formed by volcanic activity, usually over a subduction zone where one part of the oceanic lithosphere is subducting beneath another part.

J

Jasperoid – Altered wall rock containing fine-grained silica with hematite. Common in epithermal deposits.

K

Kaolin – Kaolinite. A fine (China) clay.

Kaolinization – Alteration of a mineral (usually muscovite, biotite, or feldspar) into kaolinite. See *Argillic Alteration*.

Kimberlite – Igneous rock found in volcanic pipes. Composed of peridotite, phlogopite mica, pyrope garnets, chlorite, and carbonates, commonly brecciated, carbonated, and containing pieces of the crust and mantle. Some kimberlites contain diamonds.

Kuroko-type Deposit – Japanese term for volcanic-associated massive sulfide deposit (VMS) with copper, zinc, and lead and possibly gold and silver, formed in a back arc basin environment.

L

Lamproite – Potassium- and magnesium-rich mafic to ultramafic lamprophyre-type rock of volcanic origin. Some may contain diamonds.

Lamprophyre – Minor intrusive rock containing biotite or phlogopite mica and olivine in a groundmass of feldspars or feldspathoids.

Leakage Halos – More mobile elements such as arsenic and zinc may form halos around a mineral deposit. See *Diffusion Aureoles*.

Lithology – The structure and composition of a rock formation.

Lithosphere – Upper (solid) shell of earth comprising the crust and upper mantle. Under the ocean it is up to 1,112 miles or 180 kilometers thick, and under the craton it is at least 155 miles or 250 kilometers thick.

Lode – Mineralized body or vein, or an ore deposit.

Loess – Wind-deposited silt and dust, often forming thick deposits that cover bedrock.

Loon Dung – A Canadian slang term for swampy, featureless areas that are very muddy and wet. The land is usually covered with mosquitos and black flies in the summer. The land is also underlain by permafrost so it is poorly drained.

M

Mafic – Containing more than fifty percent ferromagnesian minerals.

Magma – A mobile silicate melt with suspended crystals and dissolved gases, formed by total or partial melting of solid crustal or mantle rocks.

Magma Chamber – Subsurface accumulation of magma.

Magnetic Declination – The angle between true north and magnetic north at a given point on the earth's surface.

Magnetometer – An instrument that measures the strength of a geomagnetic field in gammas. One-gamma sensitivity of a magnetometer is standard.

Mantle – The inner shell of the earth between the crust and the core.

Mantle Plume – A column of hot material rising to the crust from the mantle, responsible for hot spots such as in Yellowstone National Park.

Meridian – In land descriptions, a region established by a north-south line called a principle meridian and an east-west line called a base line from which land is measured by townships (north and south) and ranges (east and west). There are thirty principal meridians spaced across the United States. Alaska's land is described using five meridians: Fairbanks Meridian, Copper River Meridian, etc.

Massive Sulfide Deposit – See *Volcanic-associated Massive Sulfide Deposit (VMS)*.

Mesh – A screen with open spaces, measured in holes per inch or per centimeter. For example, thirty-mesh has thirty holes per inch. A negative eighty-mesh screens flour-like material while rejecting coarser sand-sized particles.

Mesothermal Deposit – A deposit formed at 392°F to 572°F or 200°C to 300° C and at depths of 0.75 to 2.79 miles 1.2 to 4.5 kilometers. Generally found associated with intrusive igneous rocks. Made up of fracture fillings and replacement bodies, often with zoning from high temperature to low temperature minerals according to their distance from the magma.

Metamorphic rock – Rock recrystallized from preexisting rocks by changes in temperature and pressure and by chemical action of fluids. Often foliated.

Meteoric water – Groundwater derived from rainfall or infiltration.

Mica schist – A metamorphic rock rich in muscovite mica.

Mid-ocean ridge – Accretive plate margin in an ocean basin where new oceanic lithosphere is created as plates move apart, or rift, and basalt is extruded.

Mississippi Valley-type Deposit – Carbonate-hosted base metal deposit. Named after lead-zinc deposits in the Mississippi Valley.

N

Net Smelter Return (NSR) – A percentage of the profits from smelting ore from a deposit after transportation charges and smelter fees are subtracted.

Nugget Effect – In an assay, it is a false spike in values.

O

Obduction – Process in which one plate of the lithosphere is transported onto another plate (over a subducted edge).

Oceanic Crust – Thin (about four and a half miles or seven kilometers), young upper layer of ocean floor rocks composed of sediments, pillow basalts, and a bottom layer of gabbro and ultramafic rocks.

Ophiolite – Ultramafic and mafic rocks from the oceanic crust emplaced on to continental crust by obduction. A suite of gabbro, pillow lavas, and deep-sea sediments.

Option to Purchase – An agreement between a prospector and a mining company giving the company the right to explore a claim block for an annual fee, and with a specified amount for purchasing the claims.

Oxidation – A chemical reaction involving electron transfer, often (but not always) in the presence of oxygen, resulting in different compounds. For example, rust forms when oxygen combines with iron.

Oxide Minerals – Formed by the combination of an element or elements with oxygen. Minerals formed by oxidation.

P

Paleoplacer Deposit – Ancient, cemented stream gravels containing heavy minerals such as gold. The Witwatersrand in South Africa is a famous paleoplacer gold deposit.

Pan Concentrate – Higher specific gravity rock and mineral fragments concentrated in a gold pan after stream gravels have been panned down.

Paper Claim – A mining claim recorded but not marked or posted on the ground.

Pathfinder Element – In geochemical exploration, elements that are known to be associated with certain minerals, and that disperse widely from a deposit, forming an aureole or halo. Arsenic, often associated with gold, is mobile in soil and water and is used as a pathfinder mineral for gold. See *Indicator Element*.

Paucity – Small quantities.

Pegmatite – A very coarse-grained igneous rock formed in the late stages of crystallization of the granitic melt.

Peridotite – A dark-colored, coarse-grained ultramafic igneous rock containing forty-nine percent olivine and pyroxene, but no feldspar or quartz.

Permafrost – Rock, soil, and sediment in which temperatures remain below freezing. Permanently frozen ground.

Phenocryst – A large crystal or mineral grain within a fine-grained matrix of an igneous rock. Formed during the early phase in the cooling of a magma when the magma cools slowly.

Pillow Basalt – Basalt that forms underwater when erupting lava is chilled rapidly by water. May be spherical, elliptic, or flattened pillow shapes.

Pitchblende – A brown to black radioactive, massive form of uraninite. The chief ore of uranium and radium.

Placer (Deposit) – Mineral deposit formed by the sorting or washing action of water. A glacial or alluvial deposit of sand or gravel containing eroded particles of valuable minerals.

Pleistocene – Epoch of geologic time from about 2,000,000 years ago to 10,000 years ago. Ice Age.

Pluton – A large igneous body that crystallized beneath the surface.

Porphyry Copper Deposit – A large, low-grade disseminated deposit of copper which may also contain minor molybdenum, gold, and silver. Usually in a porphyritic granite host rock.

Porphyritic – Having a texture (usually in igneous rocks) in which some crystals are larger than others.

Propylitic Alteration – Development of chlorite, epidote, clay minerals, and pyrite, from widespread effects of fluids, heat, and gases from a mineral deposit. The outermost area of alteration.

Pyrite – Iron pyrites, a brass-colored mineral; or fool's gold. Common sulfide mineral. Used to produce sulfuric acid.

Pyrope – Deep red to black magnesium garnet found in kimberlite and associated placers. A pathfinder mineral in the search for diamond-bearing kimberlite pipes.

Pyrrhotite – Common ferromagnetic sulfide mineral (magnetic pyrites).

Q

Quartz Monzonite – A granitic rock with quartz and feldspars.

R

Radiometric – Ground and airborne geophysical surveys using Geiger counters (ground only), scintillometers, and gamma ray spectrometers to locate radioactive elements.

Refractory Ore – Ore from which it is difficult to extract the valuable metal.

Replacement Ore Body – Formed by replacement of existing rocks, especially carbonate rocks. For example, skarn.

Rhyolite – Fine-grained volcanic rock composed of quartz, potassium feldspars, and plagioclase. Extrusive equivalent of a granite.

Rift – Break or split in the earth's crust. Occurs in oceanic crust at the oceanic ridges. Continental rifts are where new oceans form and the two halves become separate continents.

Roll front – In a uranium deposit, the reducing environment in a sedimentary bed where uranium is concentrated.

Roof Pendant – A large mass of country rock in the roof of an igneous intrusion.

S

Schist – Metamorphic rock with strong foliation (layers or schistosity) from parallel orientation of plate-like minerals such as mica.

Scintillometer – Instrument used in radiometric surveys for measuring alpha, beta, and gamma rays. More sensitive than a Geiger counter.

Scrap Mica (Flake Mica) – Fine-grained mica.

Scree Slopes – A mantle of rock fragments on a slope below a rock face.

Seafloor Spreading – Area where oceanic plates move apart at oceanic ridges and enlarge the oceanic basins between the continents. See *Mid-ocean Ridge.*

Secondary Enrichment (Supergene Enrichment) – Increase in the grade of an ore deposit as water percolating down from the surface carries sulfide metals into the zone of oxidation at the water table where the minerals precipitate as native metals, carbonates, silicates, and oxides.

Section – A portion of a township measuring one mile by one mile square. A township consists of thirty-six sections.

Selvage – The walls of a vein, the margin where alteration effects are found.

Sericite – Fine-grained variety of muscovite mica created from aluminum-rich rocks such as feldspars during hydrothermal alteration.

Sericitization – A type of wall rock alteration that produces sericite and quartz.

Serpentinite – Serpentine minerals with talc, iron-titanium oxide, and calcium-magnesium carbonates.

Serpentinization – Alteration of ultramafic rocks such as dunite, peridotite, and pyroxenite into serpentine.

Silicified – Altered by cementation, or mineral replacement involving quartz or cryptocrystalline silica, usually in wall rock.

Sinter (Geyserite) – A variety of opal found in and around geysers and hot springs.

Skarn – Deposit of molybdenum, tungsten, copper, lead, zinc, gold, or iron formed at high temperatures in carbonate sediments at the contact with an igneous body. Characterized by calc-silicate minerals such as garnet, pyroxene, and wollastonite.

Sourdough – A slang term for someone who has overwintered in the north.

Specific Gravity (SG) – Ratio of the mass of a substance to the mass of an equal volume of distilled water, at one atmospheric pressure and a temperature of 4° C or 39° F.

Stockwork – Closely-spaced veinlets, commonly in and around granitic to dioritic igneous intrusions.

Stratabound Mineral Deposit – Deposit restricted to a horizontal area in sedimentary bedding. Conformable with the strata. Usually deposited at the same time as the sedimentary bedding was created.

Stream Sediments – The fine sands and silts that are collected during a geochemical survey, to analyze for the elements present in the stream drainage.

Stringers – A narrow vein or irregular filament of mineral traversing a rock mass of a different material.

Subduction – Underthrusting of an oceanic plate into the mantle at a continental plate margin.

Sulfide (Sulphide) Minerals – Metal compounds with sulfur. An economically important group of ore minerals. For example, galena and sphalerite.

T

Talus – Rock debris at the foot of a cliff or steep slope.

Till – Unstratified, unsorted glacial debris of clay, sand, boulders, and gravel.

Tor – A high, rocky hill or crag.

Tourmalization – Alteration of rock by the addition of significant amounts of tourmaline. Associated with medium- to high-temperature mineralization.

Township – A government-surveyed land area measuring six miles by six miles square. Composed of thirty-six mile square sections.

Trommel Wash Plant – Cylindrical, rotating screen used in placer mining to wash the gravels and screen out larger-sized rocks.

tufaceous – Being a porous igneous rock that is usually stratified and is formed by the consolidation of volcanic ash, dust, etc.

Turbidite – Unsorted fine- to cobble-sized rock deposit composed of silts, mudstones, sands, and cobbles in a shallow to deep marine basin from sediment landslides and flows that are triggered by gravity and earthquakes.

U

Ultramafic – Igneous rock with greater than ninety percent ferromagnesian minerals, composed of olivine and pyroxene. Gabbro and peridotite are ultramafic rocks.

V

Vein – Sheet-like, flat, discordant (cutting across strata), or fracture-filling deposit (or mineralized body), usually of quartz and possibly precious and base minerals.

Volcanogenic Massive Sulfide Deposits (VMS or Volcanic-associated Massive Sulfide Deposit) – Stratiform orebody consisting mainly of iron sulfides (pyrite and/or pyrrhotite) and either predominately copper or zinc with lead. Formed by volcanic exhalative activity such as in undersea black smokers.

W

Weathering – Near-surface disintegration and decomposition of rocks and sediments by mechanical and chemical processes.

Wollastonite – A pyroxenoid formed by the metamorphism of limestones in contact with magma in a skarn. A white to gray mineral, essentially a calcium silicate.

X

Y

Z

GOVERNMENT AGENCIES AND BUREAUS LIST

*BUREAU OF LAND
MANAGEMENT LISTINGS*

BUREAU OF LAND MANAGEMENT
Street: 1620 L Street NW
Washington, DC 20036
Mailing: 1849 C Street NW
Washington, DC 20240
www.blm.gov

ARIZONA BLM STATE OFFICE
3707 North 7th Street
Phoenix, AZ 85014
(602) 650-0528
www.az.blm.gov

CALIFORNIA BLM STATE OFFICE
2800 Cottage Way, Suite W-1834
Sacramento, CA 95825-1886
(916) 978-4400
(916) 978-4699 Fax
www.ca.blm.gov

COLORADO BLM STATE OFFICE
2850 Youngfield Street
Lakewood, CO 80215-7076
(303) 239-3600
(303) 239-3933 Fax
www.co.blm.gov

IDAHO BLM STATE OFFICE
1387 South Vinnell Way
Boise, ID 83709-1657
(208) 373-4000
(208) 373-3899 Fax
www.id.blm.gov

MONTANA BLM STATE OFFICE
Street: 5001 Southgate Drive
Billings, MT 59101
Mailing: P.O.Box 36800
Billings, MT 59107-6800
(406) 896-5004
(406) 896-5298 Fax
www.mt.blm.gov

NEVADA BLM STATE OFFICE
1340 Financial Boulevard
Reno, NV 89502
(775) 861-6400
www.nv.blm.gov

NEW MEXICO BLM OFFICE
1474 Rodeo Road
Santa Fe, NM 87504
(505) 438-7502
www.nm.blm.gov

OREGON BLM STATE OFFICE
Street: 333 SW First Avenue
Portland, OR 97204
Mailing: P.O.Box 2965
Portland, OR 97208
(503) 808-6002
(503) 808-6308 Fax
www.or.blm.gov

UTAH BLM OFFICE
P.O.Box 45155
Salt Lake City, UT 84145-0155
(801) 539-4001
(801) 539-4013 Fax
www.ut.blm.gov

**WASHINGTON BLM
STATE OFFICE**
Street: 333 SW First Avenue
Portland, OR 97204
Mailing: P.O.Box 2965
Portland, OR 97208
(503) 808-6002
(503) 808-6308 Fax
www.or.blm.gov

WYOMING BLM STATE OFFICE
Street: 5353 Yellowstone
Mailing: P.O. Box 1828
Cheyenne, WY 82003
(307) 775-6256
(307) 775-6082 Fax
www.wy.blm.gov

*USDA FOREST
SERVICE LISTINGS*

USDA FOREST SERVICE
P.O.Box 96090
Washington, DC 20090-6090
(202) 720-USDA

**USDA FOREST SERVICE
INTERMOUNTAIN REGION (R-4)**
Federal Building
324 25th Street
Ogden, UT 84401-2310

**USDA FOREST SERVICE
NORTHERN REGION (R-1)**
Street: Federal Building
200 Broadway
Missoula, MT 59807-7669
Mailing: P.O.Box 7669
Missoula, MT 59807-7669

**USDA FOREST SERVICE
PACIFIC SOUTHWEST REGION (R-5)**
1323 Club Drive
Vallejo, CA 94592

**USDA FOREST SERVICE
PACIFIC SOUTHWEST
RESEARCH STATION**
Street: 800 Buchanan Street,
     West Building
Albany, CA 94710-0011
Mailing: P.O. Box 245
Berkeley, CA 94701-0245

**USDA FOREST SERVICE
PACIFIC NORTHWEST RESEARCH
STATION**
Street: 333 SW 1st Avenue
Portland, OR 97208
Mailing: P.O. Box 3890
Portland, OR 97208

**USDA FOREST SERVICE
ROCKY MOUNTAIN REGION (R-2)**
Street: 740 Simms Street
Golden, CO 80401
Mailing: P.O. Box 25127
Lakewood, CO 80225

**USDA FOREST SERVICE
ROCKY MOUNTAIN RESEARCH
STATION**
240 West Prospect Road
Fort Collins, CO 80526-2098

**USDA FOREST SERVICE
SOUTHWESTERN REGION (R-3)**
Federal Building
517 Gold Avenue SW
Albuquerque, NM 87102

*US GEOLOGICAL
SURVEY LISTINGS*

US GEOLOGICAL SURVEY
(888) ASK-USGS (275-8747)
www.usgs.gov

**US GEOLOGICAL SURVEY
WESTERN REGION**
345 Middlefield Road
Menlo Park, CA 94025
(650) 853-8300

US GEOLOGICAL SURVEY
CENTRAL REGION
Box 25046 Denver Federal Center
Denver, CO 80225
(303) 236-5900

EARTH RESOURCES
OBSERVATION SYSTEMS
DATA CENTER OF THE USGS
http://edcwww.cr.usgs.gov

*USGS STATE REPRESENTATIVES
OFFICES (General State Information)*

ARIZONA GEOLOGICAL SURVEY
520 North Park Avenue, Suite 221
Tucson, AZ 85719
(520) 670-6671
(520) 670-5592 Fax
dc_az@usgs.gov

CALIFORNIA GEOLOGICAL
SURVEY
Placer Hall
6000 J Street
Sacramento, CA 95819-6129
(916) 278-3000
(916) 278-3045 Fax
dc_ca@usgs.gov

COLORADO GEOLOGICAL
SURVEY
Building 53
Denver Federal Center
Mail Stop 415, Box 25046
lakewood, CO 80225
(303) 236-4882
(303) 236-4912 Fax
dc_co@usgs.gov

IDAHO GEOLOGICAL SURVEY
230 Collins Road
Boise, ID 83702-4520
(208) 387-1300
(208) 387-1372 Fax
dc_id@usgs.gov

MONTANA GEOLOGICAL
SURVEY
3162 Bozeman Avenue
Helena, MT 59601
(406) 457-5900
(406) 457-5990 Fax
dc_mt@usgs.gov

UTAH GEOLOGICAL SURVEY
1745 West 1700 South
Room 1016 Administrative Building
Salt Lake City, UT 84104
(801) 975-3350
(801) 975-3424 Fax
dc_ut@usgs.gov

OREGON GEOLOGICAL SURVEY
10615 SE Cherry Blossom Drive
Portland, OR 97216
(503) 251-3200
(503) 251-3470 Fax
dc_or@usgs.gov

WASHINGTON GEOLOGICAL
SURVEY
1201 Pacific Avenue, Suite 600
Tacoma, WA 98402
(253) 428-3600
(253) 428 3614 Fax
dc_wa@usgs.gov

WYOMING STATE
GEOLOGICAL SURVEY
2617 East Lincolnway, Suite B
Cheyenne, WY 82001
(307) 778-2931
(307) 778-2764 Fax
staterep_wy@mailcheyenne.cr.usgs.gov

ADDITIONAL LISTINGS

THE DIVISION OF GEOLOGICAL
AND GEOPHYSICAL SURVEYS, DGGS
794 University Avenue, Suite 200
Fairbanks, Alaska 99709-3645
(907) 451-5000
(907) 451-5050 Fax
http://www.dggs.dnr.state.ak.us/

OFFICE OF PUBLIC INFORMATION
U.S. BUREAU OF MINES, MS 1040
810 7th Street, NW
Washington, DC 20241-0002

OFFICE OF SURFACE MINING
1951 Constitution Avenue NW
Washington, DC 20240
(202) 208-2719
getinfo@osmre.gov

ARIZONA DEPARTMENT OF
MINES AND MINERAL
RESOURCES
1502 West Washington
Phoenix, AZ 85007-3210
(602) 255-3795 or (800) 446-4259
(602) 255-3777 Fax
http://www.admmr.state.az.us

THE ARIZONA
GEOLOGICAL SURVEY
416 West Congress Street, Suite 100
Tucson, Arizona 85701
(520) 770-3500 / (520) 770-3505
http://www.azgs.state.az.us/

THE ARIZONA
GEOLOGICAL SOCIETY
Publications
P.O. Box 40952
Tucson, Arizona 85717
(520) 663-5295
http://www.arizonageologicalsoc.org/

CALIFORNIA DEPARTMENT OF
CONSERVATION:CALIFORNIA
GEOLOGICAL SURVEY
801 K Street, MS 24-01
Sacramento, CA 95814
(916) 322-1080
dmglib@consrv.ca.gov
www.consrv.ca.gov/dmg

COLORADO GEOLOGICAL
SURVEY
1313 Sherman Street, Room 715
Denver, CO 80203
(303) 866-2611
(303) 866-2461 Fax
cgspubs@state.co.us
http://geosurvey.state.co.us

MONTANA BUREAU OF
MINES & GEOLOGY
1300 West Park Street
The University of Montana
Butte, MT 59701-8997
(406) 496-4167
(406) 496-4451 Fax
pubsales@mtech.edu
www.mbmg.mtech.edu

NEVADA BUREAU OF
MINES & GEOLOGY
Mail Stop 178
Reno, NV 89557-0088
(775) 784-6691
(775) 784-1709 Fax
nbmginfo@unr.edu
http://www.nbmg.unr.edu

NEW MEXICO BUREAU OF
GEOLOGY AND MINERAL
RESOURCES
801 Leroy Place
Socorro, NM 87801-4796
(505) 835-5420
(505) 835-6333 Fax
gic@gis.nmt.edu
http://geoinfo.nmt.edu

THE OREGON DEPARTMENT OF
GEOLOGY AND MINERAL
INDUSTRIES
800 NE Oregon Street, Suite 965
Portland, OR 97232
(503) 731-4100
(503) 731-4100 Fax
http://www.oregongeology.com or
http://www.oregongeology.state.or.us

**STATE OF UTAH DEPARTMENT
OF NATURAL RESOURCES:
DIVISION OF OIL, GAS AND
MINING**
Street: 1594 West North Temple,
Suite 1210
Mailing: P.O. Box 145801
Salt Lake City, UT 84114-5801
(801) 538-5340
(801) 359-3940 Fax
www.dogm.nr.state.ut.us

**WASHINGTON STATE
DEPARTMENT OF NATURAL
RESOURCES**
Street: 1111 Washington Street SE,
Room 148
Mailing: P.O. Box 47007
Olympia, WA 98504-7007
(360) 902-1450
geology@wadnr.gov
www.wa.gov/dnr/htdocs/ger/index/html

**WYOMING STATE
GEOLOGICAL SURVEY**
P.O. Box 3008
Laramie, WY 82071
(307) 766-2286
(307) 766-2605 Fax
wsgs@wsgs.uwyo.edu
www.wsgsweb.uwyo.edu

WORKS CITED

Bateman, Paul C. California Division of Mines *Special Report 47*. San Francisco: State of California, Department of Natural Resources, Division of Mines, 1956.

Boyle, Robert W. "Elemental Associations in Mineral Deposits and Indicator Elements of Interest in Geochemical Prospecting." Geological Survey of Canada *Paper 74-45*. Ottawa, Geological Survey of Canada, 1974.

Chaussier, Jean-Bernard. Introduction to *Mineral Prospecting Manual*. New York: El Sevier Publishing Co., 1987.

Coope, J. Alan. "Carlin Trend Exploration History: Discovery of the Carlin Deposit." Nevada Bureau of Mines and Geology *Special Publications 13*. Reno: Mackay School of Mines, 1991.

Cruikshank, Moses. *The Life I've Been Living*. Fairbanks: University of Alaska Press, 1986.

Eckstrand, O.R. "Canadian Mineral Deposit Types: A Geological Synopsis" Geological Survey of Canada, *Economic Geology Report 36*. Geological Survey of Canada, 1984.

Evans, Anthony M. *Introduction of Mineral Exploration*. London: Blackwell Scientific, 1995.

Evans, Anthony M. *Ore Geology and Industrial Minerals*. Oxford: Blackwell Scientific, 1996.

Foster, Lynne. *Adventuring in the California Desert*. San Francisco: Sierra Club Books, 1987.

Gaffin, Jane. *Cashing In*. Whitehorse: Nortech Services, 1980.

Gardner, Erle Stanley. Hunting *Lost Mines by Helicopter*. New York: William Morrow and Co., 1965.

Goodpaster Project Field Manual. Fairbanks: Avalon Development Company.

Gunther, C. Godfrey. *The Examination of Prospects*. New York: McGraw-Hill, 1932.

Irwin, Robert. *Profits From Penny Stocks*. New York: Franklin Watts, 1986.

Joraleman, Ira B. *Romantic Copper, Its Lure and Lore*. New York: Appleton-Century Co., 1936.

Mertie, J. B. "The Gold Pan: A Neglected Geological Tool." *Economic Geology, Volume 49*, 1954.

Mertie, J. B. "Monazite in the Granitic Rocks of the Southeastern Atlantic States—An Example of the Use of Heavy Minerals in Geologic Exploration." USGS *Professional Paper 1094*. Washington, DC: U.S. Government Printing Office, 1979.

Monaco, James and Jeannette Monaco. *Rock and Gem* magazine. Ventura: Miller Magazine, April, 1999.

Nader, Ralph and John Abbots. *The Menace of Atomic Energy*. New York: W.M. Norton & Co., 1977.

Peters, William. *Exploration and Mining Geology*. New York: John Wiley and Sons, 1978.

Roberts, Ralph. "Metallogenic Provinces and Mineral Belts in Nevada." Nevada Bureau of Mines *Report 13*. Reno: Mackay School of Mines, University of Nevada, 1966.

Solka, Paul. "Lost Gold Mine of Upper Tanana." Fairbanks: *Fairbanks Daily News-Miner*, 1994.

Theobald, Paul. 1957. "The Gold Pan as a Quantitative Geologic Tool." USGS *Bulletin 1071A*. Washington, DC: U.S. Government Printing Office, 1957.

Weinberg, Alvin. *The First Nuclear Era*. New York: American Institute of Physics, 1994.

Wilhelm, Walt. *Last Rig to Battle Mountain*. New York: William Morrow and Co., 1970.

Wiltse, Milt. "The Use of Geochemical Stream-drainage Samples for Detecting Bulk-Minable Gold Deposits in Alaska," DGGS *Report on Investigations 88-13*. 1988.

Yeend, Warren. USGS *Bulletin 1943*. Washington, DC: U.S. Government Printing Office, 1991.

Suggested Reading

Adato, Michelle. *Safety Second, the NRC and America's Power Plants.* Bloomington, Indiana University Press, 1987.

Ashley, Roger P. and Daniel R. Shawe. "Introduction to Geology and Resources of Gold and Geochemistry of Gold." USGS *Bulletin 1857A.* Washington, DC: U.S. Government Printing Office, 1988.

Ashley, Roger P. and Daniel R. Shawe. "U.S. Gold Terranes–Part I." USGS *Bulletin 1857B.* Washington, DC: U.S. Government Printing Office, 1989.

Ashley, Roger P. and Daniel R. Shawe. "Gold-Bearing Polymetallic Veins and Replacement Deposits–Part I." USGS *Bulletin 1857C.* Washington, DC: U.S. Government Printing Office, 1989.

Ashley, Roger P. and Daniel R. Shawe. "Gold Deposits in Metamorphic Rocks–Part I." USGS *Bulletin 1857D.* Washington, DC: U.S. Government Printing Office, 1989.

Ashley, Roger P. and Daniel R. Shawe. "Gold in Porphyry Copper Systems." USGS *Bulletin 1857E.* Washington, DC: U.S. Government Printing Office, 1990.

Ashley, Roger P. and Daniel R. Shawe. "Gold-Bearing Polymetallic Veins and Replacement Deposits–Part II." USGS *Bulletin 1857F.* Washington, DC: U.S. Government Printing Office, 1990.

Ashley, Roger P. and Daniel R. Shawe. "Gold in Placer Deposits." USGS *Bulletin 1857G.* Washington, DC: U.S. Government Printing Office, 1989.

Ashley, Roger P. and Daniel R. Shawe. "Epithermal Gold Deposits–Part I." USGS *Bulletin 1857H.* Washington, DC: U.S. Government Printing Office, 1991.

Ashley, Roger P. and Daniel R. Shawe. "Epithermal Gold Deposits–Part II." USGS *Bulletin 1857H.* Washington, DC: U.S. Government Printing Office, 1991.

Bateman, Paul C. "Economic Geology of the Bishop Tungsten District." California Division of Mines *Special Reports 47*. San Francisco: State of California Division of Mines, 1956.

Berg, Henry C. "Metalliferous Lode Deposits of Alaska." USGS *Bulletin 1246*. Washington, DC: Reprinted by *Alaskan Prospector*, Fairbanks, Alaska, 1967.

Blanchard, Roland. "The Interpretation of Leached Outcrops."Nevada Bureau of Mines *Bulletin 66*. Reno: University of Nevada Press, 1968.

Boyle, Robert W. "Elemental Associations in Mineral Deposits and Indicator Elements of Interest in Geochemical Prospecting." Geological Survey of Canada *Paper 75-45*. Ottawa: Geological Survey of Canada, 1974.

Boyle, Robert W. "The Geochemistry of Gold and Its Deposits" Geological Survey of Canada *Bulletin 280*. Ottawa: Geological Survey of Canada, 1979.

Bureau of Land Management. Mining Claims and Sites on Federal Lands. Washington, DC: U.S. Government Printing Office, 1991.

Cameron, Eugene N. "Pegmatite Investigations in Colorado, Wyoming and Utah 1942-44."USGS *Professional Paper 225*. Washington, DC: U.S. Government Printing Office, 1954.

Chaussier, Jean-Bernard. *Mineral Prospecting Manual*. French Bureau de Recherches Geologiques et Minieres. New York: El Sevier Publishing Co., 1987.

Cobb, Edward H. and Reuben Kachadoorian. "Index of Metallic and Nonmetallic Mineral Deposits of Alaska." USGS *Bulletin 1139*. New York: El Sevier Publishing Co., 1961.

Cobb, Edward H. "Placer Deposits of Alaska." USGS *Bulletin 1374*. Washington, DC: U.S. Government Printing Office, 1973.

Conatser, Estee. *The Sterling Legend: The Facts Behind the Lost Dutchman Mine*. Baldwin Park: Gem Guides Book Co., 1972.

Cruikshank, Moses. *The Life I've Been Living*. Fairbanks: University of Alaska Press, 1986.

Daley, Ellen. "The Guide to Alaska Geologic and Mineral Information." *Information Circular 44*. Fairbanks: State of Alaska, Division of Geological and Geophysical Surveys.

Eckstrand, O.R. "Canadian Mineral Deposit Types: A Geological Synopsis."*Economic Geology Report*. Ottawa: Geological Society of Canada, 1984.

Eichstaedt, Peter. *If You Poison Us*. Santa Fe: Red Crane Books, 1994.

Emmons, William. *Gold Deposits of the World*. New York: McGraw-Hill Book Co., 1937.

Evans, Anthony M. *Introduction of Mineral Exploration*. London: Blackwell Science, 1995.

Evans, Anthony M. *Ore Geology and Industrial Minerals*. London: Blackwell Science, 1996.

Finch, Warren et al. "Nuclear Fuels." USGS *Professional Paper 820*. Washington, DC: U.S. Government Printing Office, 1973.

Fipke, C.E., J.J. Gurney, and R.O. Moore. "Diamond Exploration Techniques Emphasizing Indicator Mineral Geochemistry and Canadian Examples."*Geological Survey of Canada Bulletin 423*. Ottawa: Geological Survey of Canada, 1995.

Force, Eric R. "The Bisbee Group of the Tombstone Hills, Southeastern Arizona–Stratigraphy, Structure, Metamorphism, and Mineralization." USGS *Bulletin 2042-B*. Washington, DC: U.S. Government Printing Office, 1996.

Foster, Lynne. *Adventuring in the California Desert*. San Francisco: Sierra Club Books, 1987.

Fraser, Hugh. *A Journey North: The Great Thompson Nickel Discovery*. Thompson, Manitoba: International Nickel Company of Canada, 1985.

Frolick, Vernon. *Fire Into Ice: Charles Fipke and the Great Diamond Hunt*. Vancouver, British Columbia: Raincoast Books.

Gaffin, Jane. *Cashing In*. Whitehorse: Nortech Services, 1980.

Gardner, Erle Stanley. *Hunting Lost Mines By Helicopter*. New York: William Morrow and Company, 1965.

Glover, T.E. *The Lost Dutchman Mine of Jacob Waltz–Part I: The Golden Dream*. Phoenix: Cowboy Miner Productions, 1998.

Glover, T.E and George "Brownie" Holmes. *The Lost Dutchman Mine of Jacob Waltz–Part II: The Holmes Manuscript*. Phoenix: Cowboy Miner Productions, 2000.

Goodpaster Project Field Manual.

Gunther, C. Godfrey and Russell C. Fleming. *The Examination of Prospects.* New York: McGraw-Hill, 1932.

Hanley, J.B. "Pegmatite Investigations in Colorado, Wyoming and Utah 1942-44." USGS *Professional Paper 227.* Washington, DC: U.S. Government Printing Office, 1950.

Hart, Matthew. *Golden Giant: Hemlo and the Rush for Canada's Gold.* Vancouver: Douglas and McIntyre, 1985.

Hawks, H.E. "Principles of Geochemical Prospecting." USGS *Bulletin 1000-F.* Washington, DC: U.S. Government Printing Office, 1957.

Hawkes, Herbert E., Arthur W. Rose, and John S. Webb. *Geochemistry in Mineral Exploration.* New York: Harper and Row, 1979.

Heiner, Virginia Doyle. *Alaska's Mining History: A Source Document.* Anchorage: State of Alaska, Division of Parks, Office of History and Archaeology, 1977.

Hicks, Carol L. and Spencer R. Titley. Eds. *Geology of the Porphyry Copper Deposits Southwestern North America.* Tucson: University of Arizona Press, 1966.

Hutchinson, Charles S. *Economic Deposits and Their Tectonic Setting.* New York: John Wiley & Sons, 1983.

Hutton, James. *Theory of Earth.* New York: Hafner, 1959 Reprint.

Irwin, Robert. *Profits From Penny Stocks.* New York: Franklin Watts, 1986.

Jahns, Richard. "The Study of Pegmatites."*Economic Geology's Fiftieth Anniversary Volume,* edited by Alan Bateman. Lancaster: Lancaster Press, 1955.

Jensen, Lone. *The Copper Spike.* Anchorage: HaHa Press, 1975.

Jensen, Mead L. and Alan M. Bateman. *Economic Mineral Deposits.* New York: John Wiley and Sons, 1979.

Johnson, Albert. *Carmack of the Yukon.* Seattle: Epicenter Press, 1990.

Johnson, Thomas B. and Peter Steen. *Radioactive Waste.* Berkeley: University of California Press, 1980.

Joralemon, Ira B. *Romantic Copper, Its Lure and Lore.* New York: Appleton-Century Co., 1936.

Kurtak, Joseph. *Mine in the Sky.* Chelsea: Bookcrafters, 1998.

Kurtak, Joseph. "The Upside-Down Mine". *Rock and Gem* Magazine, (December 1997). Ventura: Miller Magazine, 1997.

Levinson, A.A. *Introduction to Exploration Geochemistry*. Wilmette: Applied Publications, 1974.

Lincoln, Francis Church. *Mining Districts and Mineral Resources of Nevada*. Las Vegas: Nevada Publications, 1923.

Lindgren, Waldemar. *Mineral Deposits*. New York: McGraw-Hill Book Co., 1928.

Lovering, T. G et al. "Jasperoid in the U.S." USGS *Professional Paper 710*. Washington, DC: U.S. Government Printing Office, 1972.

MacDiarmid, Roy A. and Charles F. Park. *Ore Deposits*. San Francisco: W. H. Freeman, 1975.

McLean, Evalyn Walsh. *Father Struck It Rich*. Boston: Little, Brown & Co., 1936.

Mertie, Evelyn. *Thirty Summers and a Winter*. Fairbanks: Mineral Industry Library, 1982.

Mertie, J.B. "The Economic Geology of Platinum Metals."*Professional Paper 630*. Washington, DC: U.S. Government Printing Office, 1969.

Mertie, J.B. "The Gold Pan: A Neglected Geological Tool."*Economic Geology, Volume 49*. Lancaster: Economic Geology Publishing Co., 1954.

Mertie, J.B. "Monzanite in the Granitic Rocks of the Southeastern Atlantic States–An Example of the Use of Heavy Minerals in Geologic Exploration." USGS *Professional Paper 1094*. Washington, DC: U.S. Government Printing Office, 1979.

Nader, Ralph and John Abbots. *The Menace of Atomic Energy*. New York: W. M. Norton & Co., 1977.

Newhouse, W.H. *Ore Deposits as Related to Structural Features*. Princeton: Princeton University Press, 1942.

Page, Lincoln R. et al. "Pegmatite Investigations 1942-1945, Black Hills." USGS *Professional Paper 247*. Washington, DC: U.S. Government Printing Office, 1953.

Paher, Stanley. *Nevada Ghost Towns and Mining Camps*. Las Vegas: Nevada Publications, 1984.

Parsons, A.B. *The Porphyry Coppers*. New York: The American Institute of Mining, Metallurgical, and Petroleum Engineers, 1933.

Parsons, A.B. *The Porphyry Coppers in 1956*. New York: The American Institute of Mining, Metallurgical and Petroleum Engineers, 1956.

Peters, William. *Exploration and Mining Geology*. New York: John Wiley & Sons, 1978.

Probert, Thomas. *Lost Mines and Buried Treasures of the West*. Berkeley: University of California Press, 1977.

Prowell, Sarah and Michael Sheridan. "Stratigraphy, Structure, and Gold Mineralization Related to Calderas in the Superstition Mountains." *Frontiers in Geology and Ore Deposits of Arizona and the Southwest*. Tucson: Arizona Geological Society, 1986.

Rickard, T.A. *A History of American Mining*. New York: McGraw-Hill Book Co., 1932.

Rinehart, Dean C and Donald C. Ross. "Economic Geology of the Casa Diablo Mountain Quadrangle." California Division of Mines *Special Report 48*. San Francisco: State of California Division of Mines, 1956.

Ringholz, Raye. *Uranium Frenzy: Boom and Bust on the Colorado Plateau*. New York: W. W. Norton, 1989.

Roberts, Ralph. "Metallogenic Provinces and Mineral Belts in Nevada." Nevada Bureau of Mines *Report 13*. Reno: Mackay School of Mines, University of Nevada, 1966.

Roberts, Ralph. "Alinements of Mining Districts in Northcentral Nevada." USGS *Professional Paper 400-B*. Washington, DC: U.S. Government Printing Office.

Rynerson, Fred. *Exploring and Mining for Gems and Gold in the West*. Healdsburg: Naturegraph Company, 1967.

Sawkins, F.J. *Mineral Deposits in Relation to Plate Tectonics*. New York: Springer-Verlag, 1990.

Sinkankas, John. *Prospecting for Gemstones and Minerals*. New York: Van Nostrand Reinhold Co., 1970.

Solka, Paul. "Lost Gold Mine of the Upper Tanana." *Fairbanks Daily News-Miner*. Fairbanks: *Fairbanks Daily News-Miner*, 1994.

Sprague, Marshall. *Money Mountain*. Boston: Little, Brown & Co., 1953.

Stanton, R. L. *Ore Petrology*. New York: McGraw-Hill Book Co., 1972.

Szumigala, Dave. *Alaska's Mineral Industry*. Fairbanks: State of Alaska, Division of Geological and Geophysical Surveys.

Taylor, Roger G. *Geology of Tin Deposits*. New York: Elsevier Scientific Publishing Company, 1979.

Theobald, Paul. "The Gold Pan as a Quantitative Geologic Tool."USGS *Bulletin 1071A*. Washington, DC: U.S. Government Printing Office, 1957.

Union of Concerned Scientists. "The Risks of Nuclear Power Reactors." Cambridge: Union of Concerned Scientists, 1977.

USGS. *Minerals Yearbook Volume I: Metals and Minerals*. Washington, DC: U.S. Government Printing Office, 2000.

Voynick, Stephen. "The Harding Pegmatite Mine." *Rock and Gem* Magazine (February 1997). Ventura: Miller Magazines, 1997.

Voynick, Stephen. *The Making of a Hardrock Miner*. Berkeley: Howell-North Books, 1978.

Weinberg, Alvin. *The First Nuclear Era*. New York: American Institute of Physics, 1994.

Wilhelm, Walt. *Last Rig to Battle Mountain*. New York, William Morrow and Company, 1970.

Wilson, E. D. et al. "Arizona Lode Gold Mines and Gold Mining." *Bulletin 137*. Tucson: Arizona Geological Society, 1934.

Wilson, E. D. et al. "Gold Placers and Placering of Arizona." *Bulletin 168*. Tucson: Arizona Geological Society, 1961.

Wiltse, Milt. "The Use of Geochemical Stream-drainage Samples for Detecting Bulk-Minable Gold Deposits in Alaska." DGGS *Report on Investigations 88-13*. Fairbanks: State of Alaska, Division of Geological and Geophysical Surveys, 1988.

Yeend, Warren. "Gold Placers of the Historical Forty Mile River Region." USGS *Bulletin 2125*: Washington, DC: U.S. Government Printing Office, 1996.

NEWSPAPERS
The Northern Miner

MAGAZINES
California Mining Journal
Economic Geology
Engineering and Mining Journal
Industrial Minerals
Rock and Gem Magazine

INDEX